A Brief History of the Harvard University Cyclotrons

Richard Wilson

Mallinckrodt Research Professor of Physics

Harvard University

Published by the Harvard University Department of Physics

Distributed by Harvard University Press
Cambridge, Massachusetts and London, England 2004

Published by the Harvard University Department of Physics

Distributed by Harvard University Press

Library of Congress Cataloging-in-Publication Data

Wilson, Richard, 1926–
 A brief history of the Harvard University cyclotrons / Richard Wilson.
   p. cm.
 Includes bibliographical references and index.
 ISBN 0-674-01460-X (alk. paper)
1. Harvard Cyclotron Laboratory – History. 2. Cyclotrons – Massachusetts –
Cambridge – History – 20th century. 3. Radiotherapy – Massachusetts –
Cambridge – History – 20th century. I. Title.
 QC787.C8W55 2004
 539.7'33—dc22

                                              2003027861

Design and Production: Digital Design Group, Newton, MA USA
Printed in Canada by Webcom Ltd, Toronto.

TABLE OF CONTENTS

PREFACE

 I first visited Harvard University and its cyclotron in January 1951. I had met Professor Norman Ramsey the week before in the New York subway (Broadway line) on the way from the American Physical Society meeting in Columbia University. The director of the laboratory, "Tex" Holt invited me to the department lunch. But not until 1954 did I seriously think about working there. I accepted a position as Assistant Professor starting in summer 1955. I had been working at a cyclotron of 150 Mev at AERE, Harwell, and had already thought about the line of research that I wanted to pursue, of which the first step was to upgrade the energy of the Harvard cyclotron and to extract the proton beam. The other Professors and the cyclotron staff readily agreed and helped with the conversion. Andy Koehler modified the oscillator, aided by a fortunate visit from Professor Mackenzie, and Professor Strauch and myself shimmed the magnet and built the regenerative extraction system. Then began eight years of active physics research. It was a happy time. The physics department and the University administration were very supportive. We were not competing with each other but with the task of unraveling the secrets of nature as best as we could.

I had expected that by 1970 the cyclotron would close and all effort be transferred to particle physics at the local Cambridge Electron Accelerator. But the medical program, started by Bill Preston and Andy Koehler, was already showing promise, and was encouraged by the persistence and dedication of Andy.

In the following pages I tell the story of the cyclotrons at Harvard as best as I can. The physics experiments themselves are well described elsewhere – in the original papers in journals such as *Physical Review*, and also in a little book I wrote in 1963, *The Nucleon-Nucleon Interaction*. As often happens in fundamental physics few, if any, of the experiments are important now; science has moved on and absorbed the information. But the medical work has more lasting implications. Although Harvard/MGH were not the first to treat patients with proton beams, we have treated more than any other facility and for a long time treated more than all others put together. This gave the physicians and surgeons solid data and experience to show the advantages of this treatment modality. The list of proton medical facilities in the world tells the story.

This does not fully describe the unique environment at the cyclotron. Even after the major program of nucleon-nucleon and nucleon-light nucleus scattering was finished, experiments continued. The list of publications shows that Professors from neighboring Universities brought their students and almost as many theses were presented to other Universities as were presented to the Harvard faculty. As exemplified by some of the talks in the 50th anniversary symposium, they were universally appreciative of the welcome they received at the cyclotron laboratory. This continued with the physicians and those coming for radiation damage studies.

Indeed, the cyclotron was a happy place to work. I tried to visit once a week – usually over a sandwich lunch. There was an extraordinary rapport with the physicians at MGH. This made for an efficient operation. This efficiency was often misunderstood by others who tried to replicate the medical treatments. The cyclotron laboratory was a place to make friendships. Friendships which have lasted all one's life even though the cyclotron itself has been dismantled and the site razed.

Richard Wilson
February 2004

FOREWORD

 The Harvard Cyclotron produced its first beam on June 3, 1949 and continued to operate until 2003 – a record breaking 54 years. This uniquely long operating life for a high energy cyclotron was due to the importance and variety of its applications and to its relatively low cost for operations and maintenance. These low costs came from clever design features and from its dedicated staff.

The initial research with the Harvard Cyclotron was for nuclear and particle physics. It concentrated on using proton beams of up to 160 Mev to study the interactions of protons with target protons, deuterons, helium and other nuclei. The cross sections for protons scattered by protons were measured at a number of different proton energies. In other experiments the angular distributions of the differential cross sections and polarizations were measured. Proton–deuteron scattering experiments gave information on the neutron-proton interaction. When a beryllium target was struck by high energy protons a beam of high energy neutrons was produced at approximately 90 Mev and these were scattered in turn by protons to study the neutron proton interaction. The Harvard Cyclotron proved to be an excellent facility for training graduate students. 30 Harvard Ph.D. theses, and as many from neighboring Universities and many physics papers were based on the Harvard Cyclotron.

While directing the design of the Harvard Cyclotron, Robert R. Wilson in 1946 wrote an important paper in which he pointed out that a high energy proton beam should be effective for cancer therapy since, near the end of its range, a proton beam produces a greater ionization density. By arranging for the cancer to be bombarded from different directions but with the ends of the ranges being at the cancer, the damage to the cancer would be maximized with minimum damage to the non-cancerous tissue. This pioneering paper led to research at the Cyclotron by doctors from Massachusetts hospitals in developing procedures for cancer therapy. Most of the Cyclotron's final years were devoted to cancer therapy. The Cyclotron continued to operate until a replacement accelerator, dedicated to cancer therapy, was in operation at the Massachusetts General Hospital.

It is fortunate that this history of the Harvard Cyclotron has been recorded while many of the participants were still alive to provide historical information.

Norman F. Ramsey

Norman Ramsey was director of the Harvard Cyclotron Laboratory from about 1948 to 1950 and was responsible for the actual construction and first operation. He later was awarded the Nobel Prize in physics for his work on molecular beams.

A Brief History of the Harvard University Cyclotrons

Introduction

This is a brief history of the two cyclotrons built at Harvard University and used between 1935 and 2002. It is a distinguished history and I, Richard Wilson, am proud to have been a part of it for 47 of these 67 years. In addition to hard copy, a web based history, which can be added to at any time, exists. (http://phys4.harvard.edu/~wilson/cyclotron/history.html) In addition there is a collection of 900 photographs of the cyclotron, its work, its staff and its place in the community, which have been scanned and are available for those who wish. Of course the Harvard University Archives has papers of many of the participants for the eager historian, and several hardware items are in Harvard's museum of scientific instruments.

This work falls naturally into four periods. The first period was that of the construction and use of the first cyclotron from 1935 to 1943 when it was dismantled and taken away for war work. The next period is the construction and initial use of the second cyclotron from 1945 to 1955. The third period starts with a major upgrade in 1955 and continues until the end of major physics research in 1968, and the fourth period is of intensive use for radiotherapy until final closure in summer 2002. Production of radioactive isotopes was an important part of the operation of the first cyclotron, but was only incidental in the second cyclotron, although the list of publications shows that it was not unimportant.

Historical Background

In the first third of the twentieth century the study of Physics at Harvard for both graduate and undergraduate students continued administratively under the Faculty of Arts and Sciences. The space occupied for study and experimentation grew with the construction of Lyman laboratory in the 1930s, one of which included a research library. The First World War had initiated the Department of Physics' role in defense. Its members had taught military personnel, served in the military, and performed defense research. The 1930s saw increased interest and investigation into the fields of nuclear science and the beginnings of computer science. In order to meet the research needs of its faculty, the Physics Department oversaw construction of a particle accelerator – a cyclotron.

The cyclotron had been invented in Berkeley, California in 1929 by Ernest Lawrence and constructed by Lawrence and his graduate student M. Stanley Livingston. Although the first nuclear disintegration experiments had been performed by Cockroft and Walton in the Cavendish laboratory in Cambridge, UK, using a rectifier multiplication device which carries their name, the cyclotrons proved to be very useful in the 1930s in nuclear disintegration experiments, and following the discovery of artificial radioactivity in 1934 by Joliot-Curie, were used widely in producing a variety of radioactive nuclei. Some of these radioactive nuclei were of interest in astrophysics, some of interest in the study of nuclei themselves and some were useful in nuclear medicine – both in diagnosis and in treatment. It seemed that every major university should have a cyclotron and indeed they were built at a number of places – Princeton, MIT (built by M. Stanley Livingston), Cornell (built by Stanley Livingston), Rochester (built by S.N. Van Voorhis and Lee Dubridge) and at Yale (built by E. C. Pollard and H. L. Schultz).

A Brief Timeline

1937   First cyclotron built at Harvard University for nuclear physics research

1943   First cyclotron dismantled and sent to Los Alamos

1948   Present synchrocyclotron built with funds from the Office of Naval Research (ONR)

1949   June 3: First 90 MeV proton beam

1956   Reconstruction at HCL – 160 MeV external beam

1961   May 25: First patient treated at HCL – neurosurgical irradiation

1963   Medical annex and treatment room #1 built with NASA funding

1964   100th patient treated

1966   Treatment charges accepted by Blue Cross/Blue Shield for neurosurgical irradiation

1967   End of Office of Naval Research funding

1971   NCI funds to MEEI and HCL to develop eye treatments

1972   Investigation started on feasibility of proton radiography

1972   Grant obtained from RANN program of NSF for the application of proton radiation to medical problems

1973   Studies on potential of using proton activation analysis to determine the calcium content of bone funded by RANN program of NSF

1974   Treatment of first patient with large proton field (11x14 cm)

1975   Treatment of first patient for intraocular malignant melanoma

1977   Treatment room #2 built with NCI and Harvard University funding

1977   1000th patient treated

1979   Eye treatment charges accepted by Blue Cross / Blue Shield

1981   Design study for new proton medical facility

1982   2000th patient treated

1985   3000th patient treated

1986   Design studies for proton beam delivery systems

1987   Treatment charges accepted by Blue Cross/Blue Shield for chordomas and chondrosarcomas

1987   4000th patient treated

1989   40th Anniversary of first Harvard proton beam

1990   5000th patient treated

1991   First patient treated in second neurosurgical irradiation program (STAR)

1993   6000th patient treated

1995   Ground Breaking, Northeast Proton Therapy Center (NPTC), Boston

1997   7000th patient treated

1999   8000th patient treated

2001   9000th patient treated

2001   First patient treated at NPTC (November)

2002   April 10 – Last treatment, 9116 patients, treated at HCL

2002   Sunday, June 2: Cyclotron High-Voltage disconnected by administrative fiat

2002   Monday, June 3: Cyclotron vacuum, cooling, fans shut down

2002   Sunday, June 30: Harvard Cyclotron Laboratory closed

THE FIRST HARVARD UNIVERSITY CYCLOTRON

Harvard faculty began thinking about a cyclotron as early as 1935. It was to be built as a joint project between the Graduate School of Engineering, (now replaced by the Division of Engineering and Applied Physics) with Professor Harry Mimno representing Electrical Engineering, and Associate Professors Kenneth Bainbridge and Jabez C. Street representing the physics department. Edward M. Purcell (later Nobel Laureate for Nuclear Magnetic resonance) was awarded the PhD in 1938 for a thesis on "The Focusing of Charged Particles by a Spherical Condenser." He became a Faculty Instructor in Physics, what was then the new title for what is now Assistant Professor, a five year term rank. After his war work in the MIT radiation laboratory he was quickly "snapped up" by Harvard with a tenure appointment. He retired as University professor and a Nobel Laureate.

In 1936 the construction of the cyclotron began in the (old) Gordon McKay laboratory, a wooden World War I building on the east side of Oxford Street. The magnet weighed 85 tons and had a 41 inch diameter pole tip. It accelerated protons up to energy of 12 Mev. In the 1960s a new Engineering Science building was built on the southern part of the Gordon McKay laboratory and the northern part was dismantled as a fire hazard in 1965. In 2002 a new building was finished in its location to house various administrative offices.

By 1938 the cyclotron construction was complete and a photograph on page five shows Professor Bainbridge, left, posing with Professor Street, right, and a graduate student Dr R. W. Hickman (kneeling).

The small control room is shown in another photograph. Dr Hickman wrote his PhD thesis on the Franck-Hertz experiment. By 1943 he was Lecturer on Physics and Communication Engineering, Assistant Director of the Physics Laboratories (under T. L. Lyman) and Assistant Director of the wartime Radio Research Laboratory (under

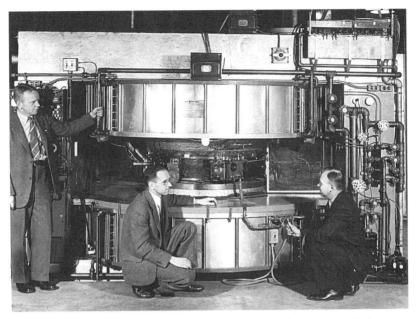

By 1938 the cyclotron construction was complete and this photograph shows Professor Bainbridge, left, posing with Professor Street, right, and a graduate student Dr R. W. Hickman (kneeling).

F. L. Terman from Stanford). Later he became Director of the Physics Laboratories until his retirement about 1968.

The cyclotron had an external beam which slowed and stopped as it passed through the air. This gives a dramatic picture of the ioniza-

Original control panel.

A technician handling one of the sources as it came out of the cyclotron.

tion of the air. The external beam was used for producing radioactive isotopes for medical purposes. A photograph shows a technician handling one of the sources as it came out of the cyclotron.

The report of the physics department to the university in 1939 states that radioactive materials were supplied to Harvard Medical School, New York Memorial Hospital and Massachusetts General Hospital in addition to uses for physics at Woods Hole Meteorological Station, MIT physics department and members of Williams College and Purdue University. It supported the work of 14 researchers in Harvard departments. Interestingly, there seemed to be no interest from the graduate school of engineering after the initial construction. This author has failed to find many references of work in this period, although in 1940 to 1941 the physics department reported that the cyclotron had been in operation for over 1,000 hours. But the end of this period and of the first Harvard Cyclotron was near.

On September 3, 1939 Great Britain and France declared war on Nazi Germany and after the Japanese attack on Pearl Harbor in December 1941 the United States joined World War II. As in World War I, many members of the Harvard physics faculty served the war effort in various ways. Some faculty members, including Professor K.T Bainbridge, had been called in 1940 to help develop radar at the radiation laboratory at MIT by E. O. Lawrence on behalf of the NDRC. But in 1943 after the establishment of Los Alamos Laboratory that Professor Bainbridge was recruited away to work on the Manhattan Project of the U.S. Army, at Los Alamos, New Mexico; He joined a highly secret team assembled by Robert Oppenheimer, to work on the development of the first atomic bomb. While there it became clear to him and to others, that a cyclotron was needed to measure various nuclear reaction cross sections of interest, which would supplement the work already being ably carried out at the Princeton cyclotron. Discussions began at a high administrative, and top secret level, between Harvard President James B. Conant (then away from Cambridge) and General Groves. It was agreed that Harvard would sell the cyclotron to the US government for $1 with an informal promise to send it back or to replace it when the war was over. It appears that Paul Buck, then Provost of Harvard University was not informed of these discussions and he later reported informally how much he agonized over making the decision to send the cyclotron.

The young scientist Dr. Robert R. Wilson was sent to Harvard to negotiate the purchase and arrange the transfer. Since the atomic bomb project was top secret, the purpose of the purchase had to be disguised from those not cleared for secret information. A medical physicist, Dr Hymer Friedell, accompanied Robert Wilson. The "cover story" was that the cyclotron was needed for medical treatment of military personnel. It was sent to St. Louis to be forwarded to an "unknown destination" (Los Alamos). Robert Wilson oversaw the shipment and Dr Hymer Friedell discusses this story in an oral history on record with US DOE. The late Professor John W. DeWire of Cornell told the author and others of being dispatched from Los Alamos to Cambridge where he took up residence whilst overseeing the dismantlement and shipping of the cyclotron to Los Alamos via St. Louis.

From the files we show a photograph of Robert Wilson (center) discussing the issue with chariman of the Cyclotron Committee of the physics department Percy Bridgman (right) with another man, believed but not confirmed to be, the late Hymer Friedell on the left. Although we can find no contemporary account of exactly what was said at the meeting, Bob Wilson, who was well known for dramatic (but essentially accurate) summaries, said 30 years later that Bridgman's response was "if you want it for what you say you want it for you can't have it. If you want it for what I think you want it for, of course you can have it."

At the time of this writing the source and amount of funds for this first cyclotron is being researched. My memory from discussion with

Robert Wilson (center) discussing the issue of giving the cyclotron to Los Alamos Laboratory with chairman of the Cyclotron Committee of the physics department Percy Bridgman (right) with another man, believed but not confirmed to be, the late Hymer Friedell on the left.

the late Roger Hickman is that the construction cost was about $40,000 of which about $20,000 came from the Rockefeller Foundation which then funded medical research.

THE SECOND HARVARD CYCLOTRON 1945–1955

Immediately following World War II, a new cyclotron and nuclear laboratory were planned. Professor Bainbridge, still at Los Alamos in the fall of 1945, wrote several letters to colleagues at Harvard to plan for a new building instead of using the old Gordon McKay laboratory. The letters show that he was, at first, unsure whether the old cyclotron would be returned or a new cyclotron would be built. Wasting no time, in 1945 Harvard University appropriated a sum of $425,000 to expand research facilities in Nuclear Physics. However, this amount was not enough to fund the construction of both a new cyclotron and a new laboratory. The U.S. Navy began a program of funding a program in basic science and through its Office of Naval Research (later a joint program of ONR and AEC administered by ONR) this department of the US government fulfilled the unwritten obligation of 1943 and offered the funding for the construction of a 700-ton cyclotron. Harvard provided the funds for the construction of a building to house both the cyclotron and a connecting laboratory. The building was originally called the Nuclear Laboratory and other nuclear facilities such as a betatron were contemplated.

We divide the history here into three phases. The first initial phase encompasses the design and initial construction, operation at 90 Mev and the research up until 1955. The second phase began in 1955 when the energy was raised to 165 Mev and the work done on nuclear physics for the next 12 years. We define a third phase as the 35 years from 1967 to 2002; years during which time the primary work was radiation therapy.

Initially, the driving force for the new cyclotron was Professor Kenneth Bainbridge. The photograph here was taken in 1960.

He persuaded Robert (Bob) R. Wilson to join the Harvard faculty as Associate Professor of Physics, after his departure from Los Alamos in summer 1946. He was to head up the team for the machine design and construction. By agreement, he was to spend the year on leave at Berkeley working with staff there on cyclotron design while Ken Bainbridge was to keep things going at Cambridge. In 1947, Bob came to Cambridge but only spent 6 months before taking up a new post as Professor of Physics and head of the Laboratory for Nuclear Science at Cornell University. He later commented that one of the facts that influenced him in his departure was being asked to do double teaching duty to make up for his "goofing off" for a year in Berkeley! So Ken Bainbridge took over from him officially in 1946–7 as the Director of the Cyclotron. But Bob's year had been very productive. In addition to establishing the major design parameters, Bob wrote a famous letter to the American Journal of Radiology [Wilson, R.R., "Radiological use of fast protons," Radiology, 47:487–491 (1946)] which presaged the later medical work. He stated that he was motivated to give some time to this medical application as "atonement for involvement in the development of the bomb at Los Alamos."

Professor Kenneth Bainbridge.

At a conference in Cambridge, UK in September 1946, which was attended by the author of this history, Richard Wilson, then a graduate student at Oxford, Professor Bainbridge described the plans for the new cyclotron. It was to occupy an area, formally empty but filled with trees, between the old Gordon McKay laboratory and Palfrey House on the east side of Oxford Street and the Divinity School on Divinity Avenue as shown in these pictures looking west and norythwest.. This site was soon filled by the cyclotron office building. This is shown in this photograph of the buildings taken, looking eastwards, from the old Gordon MacKay laboratory, and the cyclotron vault is shown under construction.

As Professor Bainbridge mischievously said, the planned neutron beam would head straight for the divinity school supposedly sending the occupants to the heavens prematurely. At a group meeting Mr. (later Dr) David Bodansky remembers an emphatic statement by Professor Bainbridge. Referring to the proposed medical work which was envisaged to be merely the production of radioactive isotopes, Bainbridge declared "There will no rats running around THIS cyclotron." Such

New cyclotron building site (looking West from Harvard Divinity School).

New cyclotron under construction (from South).

Assembly of cyclotron magnet.

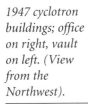

1947 cyclotron buildings; office on right, vault on left. (View from the Northwest).

blanket predictions are dangerous and often soon contradicted. Dr. R.B. "Tex" Holt, an Assistant Professor at Harvard had a wife who was doing medical research at one of the major Boston hospitals. She irradiated some of her animals in the area adjacent to the cyclotron soon after the first beam was obtained in 1949. But this was an isolated study, and the laboratory was free of the smell of animals until Dr. Raymond (Ray) Kjellberg performed his experiments on dogs and monkeys in the early 1960s preparatory to his pioneering neurosurgery treatments.

In 1948 Professor Norman Ramsey was recruited from Columbia University and became director of the Cyclotron Laboratory. Lee Davenport, who had the nebulous title of "Coordinator" stayed on and provided an effective transition. He was given the title of Associate

Group photo 1947: (Left to right) Walter Millett, Dick Stubbs, Rita McGrath, George Stilgrove, unknown, Fred Niemann, Elmer Rising, Bob Jackson, Sylvia Frye, Bob Hall, Barbara Goodwin, Charlie Brown, Dr. Lee Davenport.

Director (according to a written record) or Deputy Director (according to Professor Ramsey's memory). They assembled a fine staff shown in the group photograph taken in 1947 during the construction period.

The 1947–1948 year was very productive. The main components of the cyclotron were installed. The 650 ton magnet iron had been fabricated in Pittsburgh, PA, and machined at the local Watertown Arsenal. It was 23 ft long, 15.5 ft high and 10 ft wide. The magnet was moved

Dec. 24, 1947 delivery of part of the magnet in snow.

Photograph of the complete magnet taken on December 31, 1947. One of a set that shows the magnet assembly by Albert (Pop) Poperell with his special crew from Bigge Drayage Co. of California.

in 14 separate sections on the 3rd or 4th December 1947. The magnet was rigged into place by a special crew of riggers from California who had done much of the rigging for the cyclotron and other accelerators there. Part of it ws delivered in the snow.

The photograph of the complete magnet taken on December 31st 1947 one of a set that shows the magnet assembly by Albert (Pop) Poperell with his special crew from Bigge Drayage Co. of California, as written up in the Boston Globe of January 11th 1948.

The magnet coils, each weighing 37 tons, of which 30 tons was copper, were wound in the General Electric coil winding shop in Pittsfield, MA and were the largest coils (14 ft diameter) that could be shipped on the Boston and Maine Railroad to North Cambridge. Even then they could not come on the direct Boston and Albany mainline because of inadequate clearances. It was the clearance on this railroad that was the fi-

Train delivering magnet coils.

Dick Wharton, who stayed with the cyclotron all his working life, working on "dees" installed inside the magnet and mounted on a stub.

nal arbiter of the cyclotron energy! When it became time for the coils to be shipped from Springfield, GE wanted a responsible Harvard person to "collect" them. It was arranged that the chairman of the Harvard Physics Department would undertake this task. The chairman, Professor John H. Van Vleck, was a railroad buff from his boyhood and gladly agreed, provided that he could ride on the train! Mr. W.A. Williams, head of GE Power Transformer division accompanied the train with the first coil, and "Van", with Harvard engineer Frank B. Robie accompanied the second coil. From the vantage point of a freight car just behind "Van" took this photograph. The coil is in the white box on the car just behind the engine.

From then on construction proceeded rapidly. The cyclotron "dees" were installed inside the magnet and mounted on a stub shown being adjusted by the young Dick Wharton who stayed with the cyclotron all his working life. Another shows an unidentified technician adjusting the oscillator.

The logbook, a page of which is shown here, shows that on June 3rd 1949 at 2:03 in the morning, the first beam was obtained. Present were Norman F. Ramsey, Al. J. Pote, Robert (Bob) Mack, G.P.W. (unknown), Peter Van Heerden, and Lee L. Davenport. At the celebratory party the champagne cork made a dent in the ceiling plaster board. This dent was carefully preserved until an unfortunate redecoration sometime about 1980 destroyed the historical depression. On June 10, just before

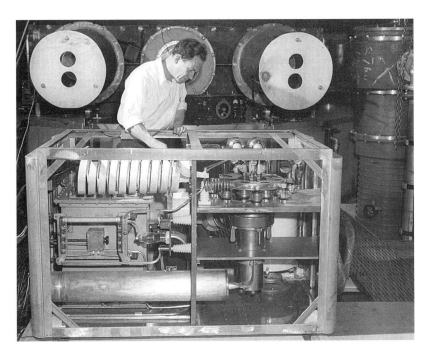

Unidentified technician adjusting the oscillator.

In 1949, Associate Director Lee Davenport (l) and Professor Ramsey (r) posed for newspapers leaning on the oscillator cabinet in front of the magnet.

the dedication, Professor Ramsey and Associate Director Lee Davenport posed for the newspapers leaning on the oscillator cabinet in front of the magnet.

The cyclotron control room. Lee L. Davenport (l) and Leo Lavetelli (r).

Later photographs show how little the control room changed over the years. Provost Paul Buck was chairman of the dedication on June 15th 1949. There was a distinguished set of speakers at either the dedication or the subsequent dinner at the Harvard Club. In addition to Norman F. Ramsey, and Lee L. Davenport, were Captain A.L. Pleasant, ONR, (Boston), Alan T. Waterman (ONR Washington), Dr Urner Lidell, and H.M. Macneille, Division of Research A.E.C. The occasion was written up in all newspapers including the NEW York Times.

In the late 1940s the enthusiasm of scientists for their research was great and that of Davenport and Ramsey was no exception. One day, after a formal dinner with the President of Harvard they returned to the cyclotron, in their dinner jackets, to find a leak using a new helium leak detector that had been delivered that afternoon. Alas no one else

A logbook page shows that on June 3rd 1949 at 2:03 in the morning, the first beam was obtained.

Mr. and Mrs. Robert Birge looking at the counters on which their data was recorded.

was present to take a photograph to record the event. A chart shows the staff during this construction period, and a photograph shows many of the staff. Many stories of this period were told at the 50th anniversary celebration by Norman Ramsey and Lee Davenport (see Appendix). The beam for the next 6 years was not at the full design energy but at a reduced energy of 110 Mev, sometimes as low as 95 Mev, because of a (temporary) failure to make the oscillator work over the full frequency range and the lack of need for immediate work using the a higher energy. Professor Ramsey, desirous of pursuing active research work at the cyclotron and even more productive work on molecular beams, for which he later won the 1989 Nobel prize in physics, arranged for Dr. R.B. Holt (Harvard PhD 1947) to become the director of the cyclotron from 1950 to 1952.

Several first rate students obtained their PhD from work at the cyclotron at this time. David Bodansky, Norton Hintz and Robert Birge were the first. The photograph shows two of them, Robert Birge and Ann (then Chamberlain) Birge, looking at the counters on which their data was recorded. At the top of the equipment rack are two binary scalers (counters) based upon the 25 year old Eccles-Jordan circuit, modified by E.B. Lewis at Cambridge in 1935 for nuclear applications, and further developed at Los Alamos by Elmore and Sands. The student had to note the lamp which showed the state of each binary in this 64 fold scaler, and perform, by slide rule, the appropriate calculations. Dr Robert Birge, son of the University of California Physics Professor Raymond Birge, was destined later to become a senior research fellow himself at the University of California at Berkeley, and Ann Birge, to become a Professor at Hayward College in California. Other students include a South African, Dr Peter Hillman, who later became a biology Professor in Hebrew University in Jerusalem.

Nikolaas Bloembergen, then a junior fellow in the Society of Fellows, also tried his hand at using the cyclotron. He, together with Peter van Heerden, measured range-energy relationships using the internal cyclotron beam and compared them to theory. But Nicholaas was to move on to a tenure position on the Harvard Faculty and to win the 1981 Nobel Prize in physics with his paramagnetic maser and his research on non linear optics. In 1950 Professor Karl Strauch joined Harvard, first as a Junior Fellow then as Assistant Professor starting in 1953. He worked tirelessly with the cyclotron for the next 10 years, before moving on to experiments with higher energy accelerators. Shortly there after Walter Selove was appointed Assistant Professor before moving on in 1956 to the University of Pennsylvania.

The ONR nuclear research contract, of which the cyclotron was the largest part, was the largest – and at first the only – government contract in the physics department. As a consequence the cyclotron laboratory became an employer of graduate students, even of those whose thesis work would be elsewhere. Two obtained their PhD before the cyclotron operated. William Cross worked on "The Conservation of Energy and Momentum in Compton Scattering (PhD 1950) and Leo Lavatelli on "Photoelectric Absorption" in 1951. Harold I. Ewen was also awarded the PhD in 1951. Ewen, with Professor Purcell, used an antenna outside the south face of Lyman Laboratory to measure "Radiation from Galactic Hydrogen at 1420 Megacycles per Second" a direct proof of the existence of interstellar hydrogen (this antenna is in the Smithsonian). Another was Paul Martin, who was awarded the PhD in 1954 for a thesis on "Bound State Problems in Electrodynamics" and who later became Dean of Applied Sciences. He remembers working in the cyclotron laboratory's electronic shop. Other non-cyclotron guests were also welcomed. In 1955–1956 Harold Furth was building pulsed high field magnets before high field superconductors were known – but he was awarded the PhD in 1960 for a thesis on "Magnetic Analysis of K- interactions in nuclei".

Space for research was scarce so in 1951–2 the nuclear laboratory building was extended to the north side to make room for an expanded machine shop and a few offices. Other appointments of note at this time included Andreas M. (Andy) Koehler, whose photograph taken at that time appears here, who was appointed at the cyclotron in some capacity that no one remembers, and which capacity Andy very quickly outgrew, and William (Bill) Preston (Ph.D. Harvard 1936) who remained as director for 20 years. At the memorial service for Bill,

.

Richard Wilson gave a eulogy outlining his work as a scientific administrator.

THE SECOND HARVARD CYCLOTRON 1955–1967

By 1953 it was already apparent that the energy of 95 Mev was too low for a long term program of nuclear and particle physics. The π meson mass had been determined to be 137 Mev, and to produce π mesons in appreciable numbers there needed to be an energy of 300 Mev or more. In addition, measurements at other cyclotrons (Rochester, Harwell, and Chicago) had shown that protons become polarized by scattering from nuclei and nucleons at energies of 130 Mev and above, but at 90 Mev the polarization is low. At the time this was merely an empirical observation, but it can be explained by noting that a nucleon of energy about 70–90 Mev suffers a phase change of 180 degrees as it passes through a heavy nucleus making the nucleus appear to be opaque (in atomic physics this is the Townsend-Ramsauer effect). In 1955 for example, Professor Mme Joliot-Curie increased the planned energy of the cyclotron being built at Orsay near Paris for this reason. Before 1953 the way of obtaining an external proton beam was by scattering from an internal target, with a consequent large loss of intensity. But in 1953 a scheme was proposed by James Tuck and Lee Teng to extract the proton beam from the Chicago cyclotron by a regenerative oscillation scheme. The theory of this process was expanded by Le Couteur in Liverpool and used to extract the beam

Andreas M. Koehler in 1954.

William (Bill) Preston, Cyclotron director for 20 years.

from the Liverpool cyclotron in 1954. In September 1955 it was decided, therefore, to rebuild the Harvard cyclotron. This rebuild coincided with the arrival at Harvard of Richard Wilson, the present historian, as Assistant Professor of Physics. Several steps were taken simultaneously:

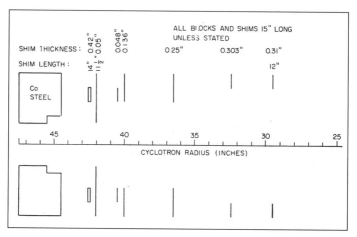

Schematic arrangement of regenerator blocks and shims to produce the field.

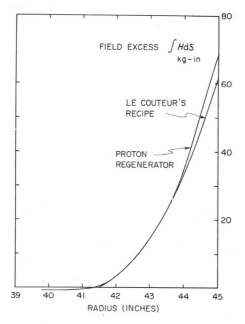

The regenerator strength versus cyclotron radius. The experimental curve is compared with LeCouteur's recipe.

The magnetic channel (top) and the regenerator (bottom) shown after dismantlement of the cyclotron in 2002.

1) The magnet was shimmed to allow cyclotron operation to a higher energy of 165 Mev.

2) The RF oscillator was adjusted so that it would oscillate over the full range of frequencies necessary.

3) A beam extraction system of the LeCouteur design was constructed.

As the beam accelerated and occupied a larger diameter orbit in the cyclotron, the protons entered a regenerator consisting of two pieces of high saturation iron, one above and below the orbiting protons at one azimuth. This is also shown schematically in regenerator schematic photograph. The regenerator was adjusted to provide an increase of magnetic field with radius that was close to Le Couteur's recipe as shown in the regenerator schematic diagram.

Shims were placed at a smaller radius to compensate for an otherwise incorrect field profile at smaller radius. An oscillation was set up between the fall off the main magnetic field and the localized increased field of the regenerator. Then the particles entered an extraction channel located at the maximum of the oscillation, at an azimuth just before the regenerator.

These photographs were taken after dismantlement of the cyclotron in 2002 but the regenerator was exactly the same as it was installed in 1956 – fourty seven years before. The rebuild had a feature unique to Harvard. It was realized that particles in the regenerator-field fall off oscillation would all have the same energy in contrast to the distribution of energies of protons striking a target under ordinary conditions. Two regenerators were constructed. One, together with the extraction channel, was used to extract the beam completely, and the other to make the monochromatic beam strike a carbon target at the other side of the cyclotron from which target scattered, polarized, protons were brought out for experiments. This is illustrated in the drawing "regenerator illustration." Which experimental program was in progress depended upon which regenerator was inserted into the magnet.

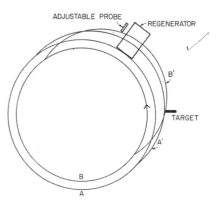

Beam extractor scheme.

The cyclotron was shut down for the rebuild in the first week of October 1955, and the beam was successfully extracted at the higher energy at the end of April 1956. The faculty and staff were, as usual for the time, enthusiastic and dedicated. For example, Professors Strauch and Wilson were shimming the main magnet until 10 p.m. on Christmas Eve. Their wives forgave them! Experiments started again within a few months. Details of the upgrade were published: "Some features of regenerative deflection and their application to the Harvard synchrocyclotron," G. Calame, P.F. Cooper, Jr., S. Engelsberg, G.L. Gerstein, A.M. Koehler, A. Kuckes, J.W. Meadows, K.Strauch and R. Wilson, Nucl. Instr. 1, 169 (1957). This paper includes all the above schematics. Subsequent improvements included a modification of the rotating condenser to adjust the frequency-time characteristics and improve the duty cycle (Koehler and LeFrançois) and a stochastic extraction scheme (Gottschalk) to improve the duty cycle still further. The photograph shows the rotating condenser with the teeth before Jacques LeFrançois made his modification to the shape of those teeth.

Over the next 10 years a number of physics experiments were performed. In addition to the persons mentioned, other faculty and research fellows who worked at the cyclotron in this period include: visiting scientist Allan Cormack (later to receive the Nobel Prize in medicine – see below), Assistant Professor Douglas Miller, Research Fellow David Measday. Dr Palmieri became Assistant Professor for a few years and Dr. Gottschalk became Professor at Northeastern University while still using the cyclotron.

The first set of experiments was a systematic study of nucleon – nucleon scattering at the energy of 160 Mev (p-p) and 135 Mev (n-p). The set included measurements of the cross section, the polarization on scattering, the depolarization on scattering, and the rotation of the plane of polarization both in and out of the plane. These experiments were described in Ph.D. theses of Palmieri, Wang, Thorndike, Hee, LeFrancois, Hoffmann, Hobbie, and published in several published papers. These experiments enabled a full phase shift analysis of the nucleon-nucleon interaction to be performed at this energy, and fit to be made to potential models. Taken together with analyses at higher energies, these showed that the spin orbit interaction was of shorter range than the rest of the interaction – deciding between the rival models of two groups of theorists, Signell and Marshak on the one hand and Gammel and Thaler on the other (the model of the second group was the correct one). The fact was later described by detailed models.

Rotating condenser with teeth (before Jacques LeFrançois made his modification to the shape of those teeth).

This work on nucleon-nucleon scattering was discussed in a small book *Nucleon-Nucleon Scattering* (Wiley-Interscience) by Richard Wilson published in 1965.

Typically the cyclotron was operated by the scientists performing the experiment and at first only he or she would be present on a night shift. Later it became clear that a second person was important for safety: the experimenter could fall down, drop a lead brick onto his toe, or otherwise get hurt. The shift change was a typical time to discuss data. On one Sunday morning Dr Allan Cormack had been on night shift, Professor Norman Ramsey was coming on day shift, and Professor Richard Wilson came by to discuss the data. But priorities changed when it was noted that the beam had disappeared, and the magnet current had gone up too high. The magnet current was regulated by comparing the voltage across a shunt with a reference, and amplifying the difference to run a bidirectional (selsyn) motor. The motor operated a variable transformer (Variac) which controlled the DC field of the DC generator. The drive for the variac was a chain and sprocket system, with limit switches. The system had failed, the limit switches failed to work, the chain had broken and the motor was struggling against the stops. Dr Cormack and Professor Ramsey sprang into action. An instant redesign took place. An O ring was used instead of the sprocket and chain drive, and two pulleys were made, one each machined by Dr Cormack and Professor Ramsey. No limit switches were needed because the O ring could slip if the drive went too far. This system was installed within the hour, and survived for about 20 years before the motor-generator set was replaced by a rectifier system acquired surplus when the Cambridge Electron Accelerator shut down. Of the three persons present that morning both Dr Cormack and Dr Ramsey were later awarded the Nobel Prize but neither of them for their skill as a machinist, important though that was.

Assistant Professor Douglas Miller set out to use the polarized neutron beam (obtained by producing neutrons at an angle of 30 degrees from the incident protons) to study neutron proton scattering. This led to the PhD theses of Russell Hobbie, and Norman Strax. Later, this neutron beam was improved and was more monochromatic, by allowing the external proton beam to impinge on a liquid deuterium target. This was done by Dr David Measday, a research fellow recruited from Oxford University, who later went to Canada and became director of the TRIUMF laboratory. Other studies included proton-proton inelastic scattering showing collisions from deep shells (Gottschalk) small angle scattering (Steinberg), neutron cross-sections (Carpenter); deuteron pickup reactions (Cooper); p-d elastic scattering (Postma) and inelastic scattering (Kuckes). Particularly notable was the first measurement of bremsstrahlung in proton-proton collisions by Shlaer and Gottshalk.

In the period 1961–1968, the interest of the Harvard physics department faculty diminished. Professors Strauch and Street had already begun an experimental program at Brookhaven National Laboratory on the Cosmotron and the AGS. When the Cambridge Electron Accelerator began to operate, in 1962, Professor Wilson also moved most of his activities, while contuining to use the cyclotron laboratory to stage his experiments. But the cyclotron itself continued an extraordinarily active life, as attested by the number of papers that were published in the period as shown in the reference list. Dr David Measday was hired as a research fellow and continued the nucleon-nucleon program. Most interestingly, there was active use of the cyclotron by scientists from neighboring Universities. Dr Gottschalk, moving to Northeastern university set up an small but active program, Dr Hohensemser and others from Clark University in Worcester were welcomed and performe nuclear physics experiments as did Professor N.S. (Sandy) Wall from MIT. Most interestingly, however, were a few radiobiological experiments by scientists from BU, which perhaps were a harbinger of more to come.

The Cyclotron staff, led by Bill Preston and Andy Koehler, continued to be outstanding. No photograph seems to exist of all the staff together, but some photographs have been located of individual machinists, assembly staff and electronic shop staff. Most of these were transferred to work on high energy experiments at the Cambridge Electron Accelerator and elsewhere as the program shifted its focus.

As the cyclotron physics progam wound down, Dr Preston took on additional tasks. Firstly as Chairman of the physics dparment and then as Director of the Physics laboratories, taking over from Dr Hickman. The photograph (next page) shows Dr Preston and Dr Hickman in his office at that time.

PROTON RADIOTHERAPY – FIRST STEPS (1961 – 1967)

As noted earlier, the first suggestion that protons could be used usefully for radiotherapy was made by Associate Professor Robert R. Wilson of Harvard University in 1947. But this idea lay dormant for many years. It was resuscitated by Dr William (Bill) Sweet, head of the Neu-

Dr Preston meeting with Dr Hickman, Director of the Physics Laboratories.

rosurgery Department in Massachusetts General Hospital, in the 1960s. Dr Sweet recruited an able colleague, Dr. Raymond (Ray) Kjellberg, to try using protons to treat various neurological lesions. Curiously Bill

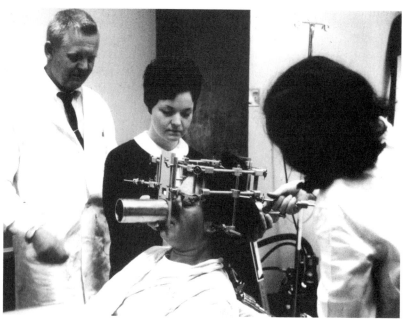

Ray Kjellberg (l) and a nurse (r) with a patient in the 1960s.

Sweet, being a trustee of Associated Universities Inc which operated Brookhaven National Laboratory, first asked Brookhaven laboratory where they could find a suitable proton beam, only to be told that there was one in his own back yard at Harvard! The Director of the cyclotron, Bill Preston, along with Andy Koehler, enthusiastically made them welcome. Not so welcome was the animal smell accompanying the first experiments on dogs and monkeys! The treatment of the first patient was described at the 2nd International Congress of Neurological Surgeons in Washington on October 17th 1961. A two year old girl was treated for a palm sized tumor, located just above the pituitary. It shrank 80%. But the improvement did not last, as the girl died within a couple of years. From then on, Dr Kjellberg decided to concentrate on diseases where removal of the pituitary gland or pituitary ablation could help. These included agromegaly and diabetic retinopathy. (In the 1980s he concentrated on arterovenous malformations). Treatment of an early patient is shown in the photograph. By 1969 Dr Kjellberg and his associate Dr Bernard Kliman had treated 46 cases of agromegaly. In 21 cases the hormone levels had dropped to normal levels.

It was opportune that the space program was just beginning and the National Aeronautics and Space Administration (NASA) was interested in the medical effects of 150 Mev protons. An energy of 150 Mev is close to the peak of the spectrum of protons that would be encountered by astronauts. The physical reason for this is the same as mentioned earlier in the discussion of the nuclear physics reasons for going to a higher energy: At an energy of 100 Mev and below, nuclei are opaque to neutrons and protons which are absorbed more readily. Any incident spectrum which has more low energy than high energy particles will have the low energy ones absorbed, leaving a peak just above the energy where the nuclei become opaque. Sensing an opportunity, a "medical annex" to the cyclotron was built using $182,000 of NASA funds. It was dedicated on November 7, 1963.

WHO WANTED THE CYCLOTRON? WHO WOULD PAY FOR IT? (1967 – 1973)

By 1967 the Office of Naval Research was closing down their basic research program. Although they continued to fund basic nuclear physics at Harvard till 1970, they would no longer provide funds for the cyclotron. The decision was faced on what to do with the facility. The cyclotron was no longer central to the physics research program of

Harvard University and the physics faculty was no longer willing to write proposals to justify a contract. It is important for the reader of this history to understand the role that physicists like to play in discoveries and development. A physicist likes to make a discovery and is delighted when it has an application that can work for the general good. But, as a physicist, he does not need to be personally involved in the development of that application, although some physicists are. Professors Pipkin, Ramsey, Strauch Street and Wilson were delighted with the continued interest, and in particular the applications to medicine, and were happy to encourage scientists from other disciplines, but wanted themselves to engage in other basic science activities.

Interestingly, as the list of publications shows, there remained a considerable interest from outside Harvard University in nuclear physics, nuclear chemistry and applications for space science. Even Ph.D. degrees were awarded at other institutions for work done at Harvard Cyclotron Laboratory – about as many in total as the number of Harvard Ph.Ds. Harvard was, and is of course, proud of the fact that it was able to help these other local institutions. But these outside scientists did not pay for the basic costs. The basic costs had to be covered by the Harvard University contract.

The most obvious choice was to close the cyclotron completely, dismantle it and use or sell the bits and pieces for other experiments or programs. The building was of great interest to the high energy physics program including Professors Ramsey, Strauch and Wilson who had ceased their experiments with the cyclotron. The exciting high energy program involved not only preparing experiments for the adjacent Cambridge Electron Accelerator, including a planned electron-positron storage ring, but also had begun to prepare experiments for Fermi lab and CERN. They were using at least half of the Cyclotron Laboratory office building and its machine shop, and in 1972 converted the basement to productive use. They had already taken over the adjacent historic Palfrey House as an office building, as well as the nearby Dunbar laboratory, released by the geology department for a computer and related work. An assembly area under a 20 ton crane was very attractive.

But the medical program of Dr Ray Kjellberg was showing great promise and it seemed improper to abandon it. Already in 1965 Dr Kjellberg discussed with Massachusetts General Hospital whether they could take over operation of the laboratory. There followed a 5 year period of discussions, irrevocable decisions, later revoked, and further discussions. These are recorded in a series of letters to various administrators about closing the cyclotron. All options were considered: moving it to another University such as Northeastern who wanted it primarily for nuclear physics experiments, in the same way the Berkeley 60 inch cyclotron moved to UC Davis, allowing another organization, hospital, University or merely different faculty, to run it in place; or closing it completely. The first decision, noted in a 1967 letter from the Director, Dr Preston, to the Harvard Dean of the Faculty of Arts and Sciences, Franklin Ford, was to close the cyclotron at the end of December 1967 when ONR support terminated. What was thought at the time to be the last "run", which was actually the first of many "last runs" in the next 33 years, occurred on February 6th 1968, with Allan Cormack (then at Tufts University) and Mildred Widgoff (of Brown University) probably on proton radiography. There was no fanfare, but Allan in his typical courteous fashion, wrote to Bill Preston to thank all connected with the cyclotron for their hospitality.

But events proceeded slowly. The always astute Harvard administrator Dick Pratt (who had started the Office of Research Contracts at Harvard University) had persuaded the Office of Naval Research in the original contract to commit the US Navy to remove the cyclotron if so requested. But they had no funds allocated for this. At their request estimates were prepared by HCL staff. These estimates of the cost for removal ranged from $191,000 to $240,000. Anxious to recoup what they could, the Navy put the cyclotron on the Excess Property List. Dr Kjellberg had some limited success in obtaining support from MGH. It was proposed to run the cyclotron one or two days a week till this problem was resolved. The conditions were described in a letter from Bill Preston to administrator Henry Murphy at MGH on July 28th 1968 and described in a letter, printed here, to Dean Franklin Ford on October 8th 1968.

But then another irrevocable decision was made to close the cyclotron in 1969 as noted in a letter by Professor Frank Pipkin. In March 1970 the US Congress, under pressure from anti Vietnam War protestors, passed the Mansfield amendment which prevented further funding by the US Navy of any work at universities. The Harvard generic "nuclear physics" contract was at an end. If the cyclotron funding had not already ceased it would.

A preferred alternative to dismantling the cyclotron and the high energy group taking the building was to "give" the cyclotron to Massachusetts General Hospital or Harvard Medical School with an informal agreement by appropriate physics department members to help in

October 9, 1968

Dean Franklin L. Ford
5 University Hall

Dear Franklin:

On at least two occasions I have announced an irrevocable decision to terminate operation of the Harvard cyclotron. The first date was December 31, 1967, and a wistful Closing Party was held at about that time to mark the end of its 20-odd years of service.

All support of cyclotron operation by the Office of Naval Research ceased at the end of 1967, but that agency was committed by its contract (wisely negotiated by Mr. Pratt!) to "restore the premises." It turns out that this will be an expensive business; an estimate by C. T. Main in February came to about $200,000, much of which would go to breaking up the large re-enforced concrete shielding blocks, too big to remove in one piece.

You will recall that for over five years the cyclotron's proton beam has been used to treat patients, with various conditions which hopefully may be arrested or improved following destruction of the pituitary gland, in a joint venture with Dr. Raymond N. Kjellberg of the Massachusetts General Hospital. It became clear early this year that the O. N. R. lacked the funds necessary to remove the cyclotron and thus that much of the space in the Cyclotron Building was not likely soon to become available for use by the Physics Department. Meanwhile, Dr. Kjellberg urged that some way be worked out so that treatment of his patients could continue. An informal arrangement was made under which we agreed to operate the cyclotron on a daily fee basis until June 30, 1968. The M. G.H. paid us from charges made to their patients. Some use of the machine was also made by NASA and by Professor Bainbridge of this Department. A special account set up for the purpose with the Comptroller was comfortably in the black on June 30, when once again I announced an irrevocable closing.

The O. N. R., in an attempt to finance removal of the cyclotron, cast about for possible buyers. No one wanted it; the best possibility seemed to be the sale to a laboratory like Brookhaven of the magnet steel at a price below the cost of scrap and insufficient to help materially. By early summer it appeared likely that our machine will be around for a long time, while the government agencies' budgets remain tight.

any problems that arose during operation. MGH would pay for its operation and pay some sort of rent for the space represented by the building. Negotiations continued about this for four more years. However, this solution ran into a two fold snag. First Dr Kjellberg was not in the center of Harvard's medical activities and second the medical community was not as aware of the possibilities as were the physicists, although Dr Milford Schulz, head of the Radiation Medicine Department at MGH supported it. Richard Wilson, the present scribbler, who was at the time Chairman of the High Energy Physics Committee of the Physics Department, wrote to, and then went to see, the Dean of Harvard Medical School, Dr Ebert, at Harvard Medical School in order to encourage HMS support for the cyclotron. But Dr Ebert found no sup-

Dean Franklin L. Ford   - 2 -   October 9, 1968

In July a significant change occurred in the official position of the M. G. H. which previously, while blessing the cyclotron medical work, refused to assume any financial responsibility for it. Partly this came about because of more widespread agreement that the proton beam irradiations were helping patients, some of whom could not otherwise be offered efficacious treatment. Negotiations were reopened and, after rather lengthy bargaining, a tentative agreement was reached on October 7. In anticipation of an agreement at an earlier date, patients had already arrived from distant parts of the country, so treatments commenced on October 8.

The Fourteen Points of agreement are enumerated under the date October 4, 1968; with them I enclose the latest pertinent correspondence between myself and Mr. Henry J. Murphy, Associate Director of the M. G. H. I enclose also copies of an exchange of letters with Dr. Padgett of the Washington O. N. R. giving permission for continued use of the cyclotron and reaffirming the Navy's responsibility ultimately to restore the premises.

On the basis of our experience in the first six months of this year, we estimate that all operating costs of the cyclotron will be recovered if we sell 40 days "time" in 12 months. Whatever slight financial risks are involved are assumed by the M. G. H; they are slight because they are free to close down operations if it is found we are running in the red. Mr. Andreas M. Koehler, Engineer and Assistant Director of the Cyclotron, and a Corporation appointee, will spend about half his effort on the project, with the M. G. H. guaranteeing a corresponding fraction of his salary. We are also offered a "credit" so as to be enabled to get started, in advance of collecting income, without our account being forced into the red.

We have budgeted for indirect costs to be collected by Harvard on sums spent here. With your permission, I will ask Mr. Janke to work this out with Mr. Murphy. I believe Harvard should recover thus at least $5,000.

We have also a provision for the payment to the Physics Department of $100 per day of operation. Even without the removal of the cyclotron and its shielding, the Harvard group working on experiments at the CEA can use several thousand square feet around the machine and in the Medical Annex. Due to the hazard of radiation, much of this space must be vacated when the machine is actually running. These payments are in partial compensation. We propose to use the money for the benefit of the CEA groups, either for renovation or remodelling of existing space or for rental of storage space in the vicinity (should this prove possible.)

I welcome your comments and hope that you will approve this agreement with the M. G. H.

Sincerely,

W. M. Preston
Director, Physics Laboratories

WMP:mb
Encs.

port in his faculty. Physicians thought that the resources of the school were better committed to finding the causes of cancer rather than treating it. Moreover, cancer experts were arguing that chemotherapy was a more promising choice for patient treatment than radiotherapy. While both these arguments seemed plausible at the time, it is now clear that they were wildly overoptimistic. Thirty six years later the causes of cancer are still elusive, and chemotherapy, by itself, is far less effective than

-3-

Dean Franklin L. Ford   October 9, 1968

P.S.   If agreeable to you, will you be so good as to inform
       Mr. Donald Porter (Office for Research Contracts, Holyoke
       Center) that you approve his requesting a no-cost extension
       of contract NAS - 9 - 8005 for one year from July 1, 1968?

       This contract was written for $18,320 by the Manned Space-
       craft Center in Houston, to buy time on the cyclotron for
       calibration of equipment for the Apollo mission. It was
       for some reason dated 05/22/68 - 06/30/68, but only $4,580
       was spent in that period. The NASA people are anxious to do
       some more work on the cyclotron as soon as it can be arranged.

when combined with radiotherapy. Rightly or wrongly, medical funding to keep the cyclotron did not seem to be forthcoming and it was decided to close the facility and make the building available by 1970. But other events intervened.

Already in 1970 there was a very marked cut back in funding for the Cambridge Electron Accelerator, leaving only a program on electron positron colliding beams, the "By Pass" program, in place. Then in 1972 the federal axe fell. The US Atomic Energy Commission (AEC) withdrew all funding starting in summer 1973. Without any known source of funds, Harvard and MIT decided to close the CEA. That would leave plenty of space for the remaining high energy physics program. Moreover, there was a reduction in the Harvard contract for high energy experiments, even though these experiments would now have to be conducted elsewhere and travel funds would be needed. The whole program was thereby reduced. There was no pressure to shut the cyclotron down if even a small amount of funding could be found.

The cyclotron was finally rescued by two important steps. Andreas (Andy) Koehler proposed a budget to the physics department showing that the cyclotron could be kept alive for one or two days a week, funded by patient fees from Dr Kjellberg's patients. It was necessary for the "someone" to "guarantee" the budget and to hold the bag if the fees failed to arrive. Since both MGH and the Medical School had showed reluctance to do this, Dr Preston and Professor Wilson persuaded the physics department to do so. A noteworthy part of the budget was that Andy Koehler, at his own insistence, went on half pay – but of course he has *never* been on half time. The second step was the arrival in Boston of Dr Herman Suit to become the new Chief of the newly reconstituted Department of Radiation Medicine at MGH. At a special Saturday morning meeting with Dr Samuel Hellman of the Joint Center for Radiotherapy at the Medical School, Dr Suit declared to Professor Wilson that one of the attractions of moving from Texas to Boston was that he could use proton therapy. Professor Wilson, slightly overplaying his hand, agreed that the physics department would keep the cyclotron open – which it did. About this time Dr Suit arranged a set of small meetings in the Boston/Cambridge area to discuss whether, indeed, protons were the best option when compared to helium or carbon ions, or negative pions. For a number of reasons the community agreed that protons were the best option. This increased the local support considerably. One of Dr Suit's first appointments was of the physicist Dr Michael Goitein, who had gained his PhD some 3 years before from Harvard Physics Department for a thesis on electron proton scattering under the guidance of Professor Richard Wilson. Dr Goitein was an expert in the use of computers. He had, as a student, put the electron scattering experimental program "on-line" and had been awarded the IBM graduate student fellowship. This interest and experience was to be put to good use in medical physics, particularly in the proton therapy program. The stage was now set for a most productive 30 year period of operation of the Harvard Cyclotron. Locally we were aided by a fortuitous circumstance. The Atomic Energy Commission had established an information exchange program with the Soviet Union. Dr John Lawrence of UC Berkeley was asked to head a 3 man team to study proton therapy in Russia. He chose to take with him to the USSR Dr Kligerman, radiotherapist from the University of Pennsylvania, and Andy Koehler. On a return visit, a three person team visited Harvard. Richard Wilson hosted a small party for all physicists and physicians involved with the HCL program – about 30 in all – and other physics department members interested in proton therapy. The Department Chairman, then Professor Paul Martin, was convinced of the importance of the program and from then on the Harvard Cyclotron could count on the support of the physics department.

Funding, however, was the most difficult task. Dr Ganz of MGH, pediatrician for Dr Kjellberg's children, suggested to Dr Charles Regan of Massachusetts Eye and Ear Infirmary (MEEI) that the proton beam was ideal for treating eye tumors and in particular the hereditary tumor retinoblastoma. Interestingly, we treated only 33 retinoblastomas, but in 2003 they are high on the list of new treatment modalities for NPTC. Dr Regan put in a proposal to the NIH but it was turned down, largely because of inadequate communication between Massachussets

Eye and Ear Infirmary and HCL. Dr Regan mistakenly described the alpha particle beam (not the proton beam) and Dr Preston felt only able to give support that cost FAS nothing. Both these defects in the proposal were remedied in a new proposal, that was finally successful, involving Dr Ian Constable and Dr Evangelos Gragoudas, both opthalmologists at MEEI. Nonetheless medical funding was slow in coming, so that the physicists Koehler, Preston and Wilson (called the Biomedical Group in the Harvard archives), started searching. On the principle of starting with the largest pocket, this small group approached the medical program of the Atomic Energy Commission which at the time was spending some $4 million a year on proton and alpha radiotherapy at Lawrence Berkeley Laboratory hoping for a small fraction – perhaps 10% of this sum. No luck. But providentially the National Science Foundation started a new program, "Research Appropriate for National Needs" (RANN). The cyclotron received two grants for this work. The first was to adapt the Harvard Cyclotron for clinical trials. The second was a pilot study of detecting calcium in the extremities of the body by proton bombardment, producing the radioactive potassium K^{38} and detecting the characteristic 2.16 Mev gamma ray. In addition, fees from the neurosurgery patients brought by Dr Kjellberg continued to arrive.

PROTON RADIOGRAPHY AND CALCIUM MEASUREMENTS

There were also other attempts to use the cyclotron for interesting medical purposes. In the late 1960s, Andy Koehler conceived the idea of using the beam for "proton radiography" – measuring the density of material rather than the high Z material characteristically shown on an x ray or a CAT scan. The aim was an early diagnosis of cancer which would manifest itself in a small density change – and therefore a change in the range of the protons. The beam has a well defined range, with the range in centimeters determined primarily by the energy and density of the material. Thus, if the density *increases* a few percent because of a tumor, the range will likewise *decrease* a few percent – and that should be measurable. This would be better than an ordinary radiograph where small density changes are difficult to observe. Ordinary X rays at the time depended upon the fact that photon absorption tends to vary as the atomic number (Z) to the 4th power. Blood, containing iron, shows up, and if barium or other high Z material is in the food, they will show up better. Proton radiography should show subtle den-

X-RAY

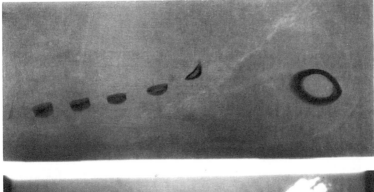

PROTON.

Proton radiography shows subtle density changes which are a sign of many lesions. The first test of the idea was dramatic, and showed itself in a famous "lamb chop" radiograph. The bottom picture, taken with protons, shows much more detail than the top picture taken with X rays.

sity changes which are a sign of many lesions. The first test of the idea was dramatic, and showed itself in a famous "lamb chop" radiograph taken by Andy and shown in many colloquia and conferences. The bottom picture, taken with protons, shows much more detail than the top picture taken with X rays.

The procedure, but strangely enough not this dramatic picture, was published in Science (see reference list). This stimulated much interest in the scientific world. In addition to Andy, Allan Cormack (by then at Tufts University) and Mildred Widgoff (from Brown University) and physician Dr V.W. Steward (from University of Chicago) worked on this idea at the Harvard Cyclotron. Allan Cormack had an even bolder approach. Could not the proton beam be used first to determine the position of the tumor and then to treat it? In this way, he hoped, an efficient, and therefore cheap, procedure could be devised for treating tumors.

This general idea was also picked up by Dr Ken Hansen, a former Harvard Student who studied it at Los Alamos National Laboratory. However, for medicine, there was no resolution of the practical matter of making it work well. Moreover, ordinary radiography and CAT scans improved considerably, and the new (Nuclear) Magnetic Resonance Imaging (MRI) was able to determine the small changes in low Z materials indicating tumors, thereby rendering proton radiography unnecessary. Maybe the idea will be resurrected at another time.

More promising, perhaps, was the idea that the production of radioactive potassium (K^{38}) by proton bombardment of calcium could lead to a bioassay for calcium. Here the idea is to locate calcium loss in the spine long before calcium loss shows up in the extremities. In a PhD thesis, (also funded by NSF in their RANN program) Dr R. Eilbert was able to find reproducibility in a phantom, made of hamburger surrounding fossil bones, of 1.5%. However medical support was not, at the time, forthcoming and the project was abandoned.

At that time (1975) we also tried to obtain funding from the National Institutes of Health for a "facility" grant, to keep the cyclotron alive for a variety of medical purposes including, of course, therapy. This, also, was unsuccessful.

PROTON RADIOTHERAPY – THE CONTINUED WORK (1975–2002)

In 1972 Dr Suit commenced a program of clinically related radiation biological experiments to assess the RBE value to be employed. These were done by Drs. I. Robertson of the Harvard School of Public Health, M. Raju of the Los Alamos Laboratory and E. Hall of Columbia University. These were in vitro studies. In parallel, a long series of RBE assays were performed on intact tissues of the laboratory mouse by Drs. M. Urano and J. Tepper. The result was that 1.10 was chosen to serve as a generic RBE value, for all dose levels and tissues.

Then in February 1974, the first patient was treated using fractionated dose radiation therapy at the equivalent of about 2 Gy/fraction (200 Rem/fraction). This patient was a boy with a posterior pelvic sarcoma. The second was a woman with a skull base sarcoma. This category of tumors now includes some 800 with really impressive results;

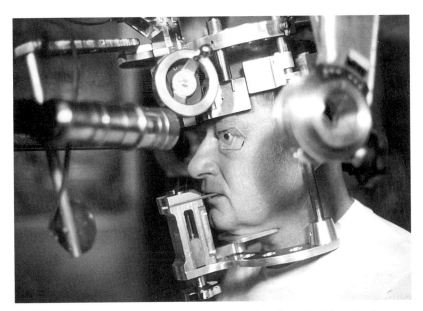

The first eye patient was Mr McKelvey from Colorado and his head is shown in this picture, holding in his mouth a deice to hold his head steady.

namely, the 10 year control results are 95% and 45% for chondrosarcoma and chordoma, respectively. The principal clinicians included Drs N. Liebisch, J. Munzenrider, M. Austin Seymour, E. Hug and H. Suit. The important clinical physicists were Drs M. Goitein, L. Verhey and A. Smith. In 1975 the first of 2,979 patients was treated for ocular melanoma by a team comprised of Drs Evanglos Gragoudas [ophthalmologic surgeon of the MEEI], John Munzenrider [radiation oncologist of the MGH] and Michael Goitein [physicist of the MGH with a Ph.D. from the Harvard Physics Department]. The first patient was Mr McKelvey from Colorado and his head is shown in this picture, holding in his mouth a deice to hold his head steady.

Dr Goitein developed the first 3D treatment planning software to be implemented in regular clinical work and used in many parts of the world. It was first designed for treatment of ocular melanoma. He also developed the concept of and brought into clinical practice: dose volume histogram (DVH), digital reconstructed radiograp (DRR), and the graphic display of uncertainty bands around isodose contours.

The year 1976 saw the start of the first NCI grant for clinical study of proton beam radiation therapy. This funding has been continuous from 1976 to the present at MGH. This grant was critical for the life of the radiation oncology program. Drs Suit and Goitein served as Co-

Principal Investigators from 1976 to 1998 when Dr Jay Loeffler at MGH became the PI.

In 1975 Dr William Preston retired from his positions as Director of the cyclotron laboratory and Director of the physics laboratories. The staff at that time included Dr Robert J. Schneider, Dr Janet Sisterson, Ms Kristen Johnson and Mr Miles Wagner in addition to Andy Koehler as Assistant Director and Bill Preston as Director. The management procedure was changed as follows: the management was vested in the acting director of the laboratory, reviewed by a management committee chaired by Professor Richard Wilson with members, Dr S.J Adelstein (Academic Dean HMS), Dr Herman Suit and Dean Richard Leahy of Harvard. This committee reported directly to the Dean of FAS and administratively bypassed the physics department. By this time the medical program at Harvard Cyclotron laboratory was well under way. There were three basic prongs.

a) Small neurosurgical (intercranial) tumors treated by the Neurosurgery Department of MGH by Dr Raymond Kjellberg and Dr Bernard Kliman, and later by Dr Chapman;

b) Eye tumors treated by Massachusetts Eye and Ear hospital by Dr Ian Constable and Dr Evangelos Gragoudas;

c) Larger tumors treated by the Radiation Medicine Department of MGH. (Dr Herman Suit, Dr Joel Tepper, Dr Michael Goitein, and Dr Lynn Verhey).

Each had its peculiarities both in funding and in treatment which differences sometimes led to stressful problems.

One of the reasons for the overall success of the program was the ability of the Harvard Cyclotron staff to maneuver independently of the rivalries, both scientific and political, between the three groups. Originally the relationship between the Cyclotron Laboratory and MGH was highly informal. By informal agreement with Dr William (Bill) Sweet, Director of the Neurosurgery Department at MGH, the Harvard Cyclotron was treated as an operating room for purposes of liability and responsibility of the medical staff. All Harvard cyclotron personnel were covered by medical malpractice insurance on the general Harvard University policy. But the increasing number of patients, and the fact there were three programs of which one, the neurosurgery program, was completely separated (on the hospital side) from the others, made a more formal agreement necessary – if only to prevent quarrelling between the physicians, surgeons and physicists. This was forced

Early staff (1979) L to R: Barbara Schnider, Janet Sisterson, Andy Koehler, William Preston, Richard Abrams, Kris Johnson, Bob Schnider III, Miles Wagner.

by a stormy interchange in 1977 about the maximum size it was safe to treat a lesion in a single fraction (see below). It was made formal and legal. The cyclotron staff also had to be made aware of the demands of patient confidentiality Harvard University negotiated a one-sided agreement. MGH was responsible for any liability arising from the treatments, but nonetheless, anyone on the cyclotron staff had the authority to decide *not* to treat a patient if he or she felt that the planned treatment was inappropriate. Fortunately such an eventuality never occurred. We collect here some photographs of the various treatments.

In the following 27 years each of these groups made major contributions, and each was in its own way essential to the whole program. However from the start the physicians at MEEI collaborated very closely with the physicians at the Radiation Medicine Department at MGH and in particular with the physicists (led by Michael Goitein) at MGH. Professor Richard Wilson went on leave and a change was made. Dr S. James Adelstein, academic dean in the medical school became Chairman of the Cyclotron management committee. The reporting was now to the Dean of Applied Sciences instead of the Dean of Faculty of Arts and Sciences. Dr Adelstein remained Chairman for the next 21 years until the final shut down in 2002.

The medical advantage of all of the treatments verified the point first raised by Robert R. Wilson in 1947. The aim of all radiation treat-

AVM : BEFORE AND AFTER

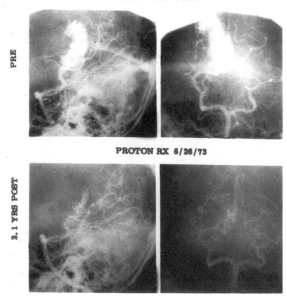

The photographs show before (top) and after X ray images (bottom) of a successful treatment for an arterovanous malformation (AVM), in which the veins in the head stuck to arteries. Two years later the malformation was gone.

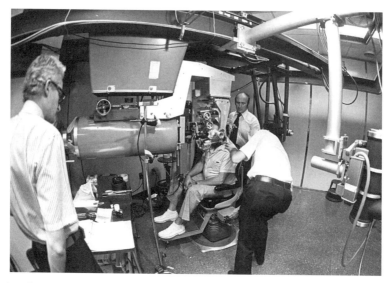

The photograph shows a typical three-way collaboration in a treatment, between Robert Schneider of HCL, Dr. Evangelos Gragoudas and Dr. Michael Goitein of MGH.

ments is to destroy cancerous and other unwanted tissue, while doing as little damage as possible to the surrounding healthy tissue. The proton beam succeeds in this for two reasons. First protons have a well defined range, with a sharp increase of ionization at the end of the range first pointed out by Sir William Bragg (the "Bragg peak"). They produce little or no damage beyond the end of the range. Secondly, protons being heavy, scatter less than the electrons commonly used for radiotherapy. If the tumor or other lesion is small, (less than 1 cm diameter) as is possible in treatments (1) and (2) it is comparatively easy to install absorbers so that the protons stop on the lesion.

The procedure adopted by Dr Kjellberg was to hold the patient's head rigidly in a stereotactic frame, and to rotate the patient, seated in a chair, about a vertical axis through the lesion. The photograph shows Dr Kjlleberg adjusting this device on an unidentified patient with an unidentified nurse looking on. Dr Kjellberg treated his patients with several beams from different (horizontal) directions, that led to an excellent dose distribution that enabled him to irradiate the pituitary gland but save the optic nerve. He varied the dose with the size of the beam in accordance with a graph he prepared of the maximum tolerated dose.

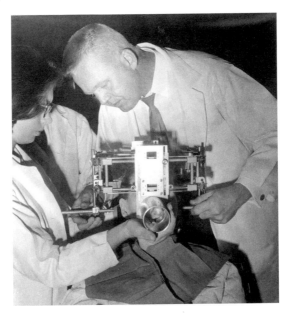

The procedure adopted by Dr Kjellberg was to hold the patient's head rigidly in a stereotactic frame, and to rotate the patient, seated in a chair, about a vertical axis through the lesion.

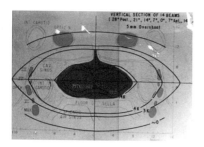

Photograph of the pituitary showing isodose curves.

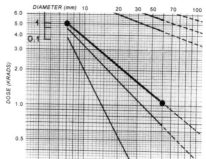

The graph used by Dr Kjellberg to decide the optimal dose. The points are experimental "safe" doses.

But the treatments were all in one day (a single "fraction"). He claimed that fractionation of the dose would not be helpful because he was trating non-cancerous lesions – a claim that others contested.

If the lesion or tumor is large, particularly as in treatments (3), the advantage of the well defined range remains but it is much harder to obtain a uniform dose distribution across the tumor. The large field arrangement was a simple one that was designed, as was so much, by Andy Koehler. Firstly the beam impinged on a scatterer to spread the beam. This resulted in a beam that was non-uniform in intensity across the beam. Then an absorber was placed in the center of this beam to absorb the higher intensity portion. Finally there was a second scatterer. This double scatterer technique is shown in the figure on the following page.

Then the range was modulated by a set of absorbers on a wheel that rotated during the treatment, allowing the proton beam to stop at various depths in the tumor.

Attached downstream of the brass aperture, which defined the lateral extent of the beam, was a plastic bolus, machined for each treatment, which fine tuned the depth of penetration of the beam into the patient. The final result, together with a comparison with other modalities, is shown in the figure on page 44.

The double scattering technique has been copied by many proton therapy facilities throughout the world.

From the beginning of this period onwards it was realized that the Harvard cyclotron was not ideal for the medical work it pioneered. Although the range of protons in tissue was 10–15 cms, this was not enough to reach all tumors from all directions. In addition, it is far

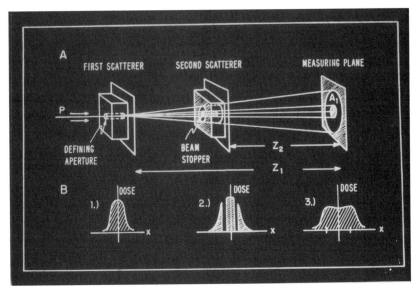

Double scatterer technique; at bottom is the dose distribution after each scatter.

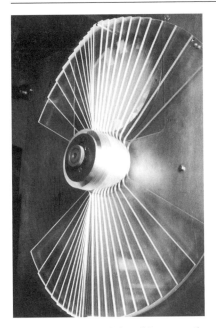

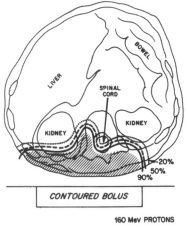

The range was modulated by a set of absorbers on a wheel that rotated during the treatment, allowing the proton beam to stop at various depths in the tumor.

Attached downstream of the brass aperture, which defined the lateral extent of the beam, was a plastic bolus, machined for each treatment, which fine tuned the depth of penetration of the beam into the patient.

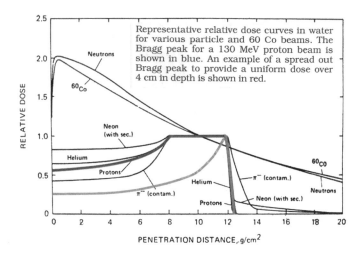

The final result, together with a comparison with other modalities.

preferable for a cyclotron to be located at the hospital. Already in 1973 Andy Koehler was thinking about small, cheaper, specialized cyclotron designs. But it was already realized that the cost of the cyclotron itself was a just small part of the total treatment cost.

Professor Bernard Gottschalk returned to the Harvard Cyclotron Laboratory as a Senior Research Fellow in 1982. One of the first tasks he undertook was to plan a new accelerator: his choice being a synchrotron because the energy is easily variable. Although attempts to obtain NIH funds for this new development failed, his design was useful in the design for the synchrotron at Loma Linda University Medical Center. That synchrotron was funded in large part by a special grant from the US Department of Energy. This grant was congressionally directed by the committee on energy in the House of Representatives chaired by Representative Lindy Boggs of Louisiana. Ms Boggs was very sensitive to the need for proton radiotherapy since her daughter, Mayor of Princeton, died of a choroidal melanoma which metastasized. Unfortunately, they became aware of our (Massachusetts Eye and Ear Infirmary, Massachusetts General Hospital and HCL) successful cures too late. We were informally asked by a committee staff member whether we would like to be included in the special appropriation, but Harvard University and MGH do not accept congressionally directed ("pork barrel") funds. A hospital based facility at MGH, would have to wait another 10 years. In 1989 the cyclotron had operated for 40 years. The staff posed for a photograph.

In 1990, after application to NIH, design funds were made available for a complete new proton therapy facility – accelerator, beam lines, treatment rooms – the lot. Professor Michael Goitein, at MGH and Harvard Medical School, was the PI of the grant and undertook the design. Construction funds were made available in 1994. The contractor for the fine building was Bechtel, and for the cyclotron and beam lines, IBA of Belgium. This became the Northeast Proton Therapy Center (NPTC) at MGH built in the exercise yard of the old Charles Street jail. The building and the first operation of the cyclotron came in on schedule, but reliable operation of the beam, beam transport and gantries was elusive. After much travail, the first patient was treated in November 2001. The whole proton therapy program began the switch to NPTC at this time, and NPTC picked up the full load from HCL by April 2002.

Staff in 1989: (L to R) Ethan Cascio, Townsed Zwart, Bob Mendelson, Gail Bradley, Dr. Bernard Gottschalk, Frank Robie, Kristen Johnson, Miles Wagner, Dr. Janet Sisterson, Sason Burns, A. Zoesman, Andreas Koehler, Eliot Hammerman, and Andrew Meglis.

Another photograph was taken of the control room — almost unchanged since 1949 — with Andy Koehler (l) and Jason Burns (r).

By 1993 Andy Koehler had been with the laboratory 40 years, many of them as acting director or director. He asked to be relieved of his duties as Director, remaining as a senior research fellow. But there was plenty of able talent. Miles Wagner took over as director and led the program for the next 9 years. In 1999 the Harvard cyclotron had been operating for 50 years. This was a record for cyclotrons. Many other cyclotrons had shut down as other studies of nuclear physics developed and high energy particle physics moved to higher energies. The staff posed in front of the cyclotron vault.

And another photograph was taken of the control panel with Jennifer Wu operating and Miles Wagner looking on.

We had already had many major parties. A "final closing" party in 1967; another "closing party" in 1970, and a 40th anniversary party in 1989. In 1999 we had to celebrate once again. We did so with a one day symposium, discussed in the Appendix, followed by a dinner at which Andy Koehler's formal retirement was announced. But with Andy, as with so many loyal Harvard people retirement did not mean stopping work.

On Wednesday April 10th 2002 the Harvard cyclotron delivered its

Cycleron staff in 1999: Above L to R: Jeffrey Bellerose, Mauricio Fernandes, Yetin Orshen, Jim Vlahakis, Miles Wagner (Director), Jenifer Wu, Diedre Foley, Alvaro Hernandes. Below L to R: Kristen Johnson, Bernard Gottschalk, Greta Ahlberg, Ethan Cascio, Elliot Hammerman, Allison Aaron Cascio, Alice Coggeshall, Andy Koehler (former Director), Gail Bradley, Xiaowei Yan.

Control panel in 1999: Miles Wagner (Director) (l); Jennifer Wu (Operator) (r).

last treatment having treated 9,116 patients. This patient was a young boy with bilateral retinoblastoma – a hereditary cancer of the eye. The photographs shows the child with his grandmother and being adjusted for treatment by Dr Mukhai, with Nurse Patricia MacManus watching.

Starting when he was 2 months old and continuing till he was 4 months old, he received 22 irradiations to each eye. We anticipate that he will be cured. A total of 2,979 eye tumors have been treated along with 3,687 neurosurgical lesions and 2,449 large tumors at various sites. A dedicated group of professors, physicians, physicists, nurses, operators and technicians from Harvard and MGH attended a small celebration of this work in the evening. The last photograph of Professor Herman Suit and Dr Munzen-reide in the control room, was taken at this celebration. Alas, attempts to have full newspaper coverage of the event were prevented by an insensitive Harvard administration. But the continued impact of the Harvard cyclotron's proton therapy program is not merely the success of the local successor (NPTC) at MGH.

On Wednesday April 10th 2002 the Harvard cyclotron delivered its last treatment having treated 9,116 patients. This patient (r) was a young boy with bilateral retinoblastoma — a hereditary cancer of the eye.

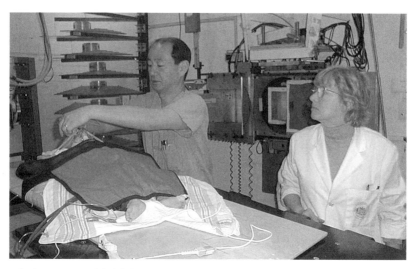

The last patient (child) being adjusted for treatment by Dr Mukhai, with Nurse Patricia MacManus watching.

It is the success of the 19 other locations where the HCL/MGH treatments have been copied or are planned. These are listed in the Table.

OTHER EXPERIMENTS – RADIATION DAMAGE STUDIES

Professor Herman Suit (r) and Dr John Munzenreide (l) in the control room; photo taken at the final celebration.

The first use of the cyclotron for radiation damage studies came when ATT needed to test their transistors in order to determine whether they would survive in space. In space there are a number of cosmic ray protons with a peak in the spectrum around 150 Mev. In 1961 a former graduate student of Professor Robert Pound, Dr Walter Brown, then at Bell Telephone Laboratories in Murray Hill, NJ, brought some of the equipment to the cyclotron to be bombarded with 150 Mev protons. The equipment survived the test, and so did the equipment on board the Telestar satellite. NASA also realized that there was a need to understand not only how equipment behaved in the radiation environment of space, but also how people behaved in this hostile envionment.

That was the primary reason that NASA funded the construction of the Medical Annex to the cyclotron. NASA also funded a special cyclotron with energy of about 500 Mev in Newport News, Virginia to perform radiation damage studies for satellite communication equipment and components. But the NASA cyclotron proved too cumbersome for the task and it was shut down in the late 1960s. Over the years, NASA itself as well as contractors for NASA, regularly brought equipment to the Harvard Cyclotron Laboratory for test. The scientists would typically have the cyclotron to themselves for the whole weekend (when medical work was not being done) with a cyclotron staff member, most recently Mr. Ethan Cascio, to help them. The list of publications and reports by the scientists from NASA and contractors shows the importance of this work – which was largely ignored by all but a few of the HCL staff.

At the time of the closing of the cyclotron, in April 2002, there was enough interest in the radiation damage studies to pay entirely for the limited operation of the cyclotron needed to perform this work. This was a very interesting development in view of the financial problems of 35 years earlier. This led to a decision which was, perhaps, the inverse of that made 35 years before. One HCL staff member, Ethan Cascio, was personally interested in keeping the cyclotron operating. Andy Koehler and Bernie Gottschalk, although at retirement age, were willing to help in a limited way. If it became known that the cyclotron was easily available – with no wait for machine time provided money was available – it is likely that other potential users would have come out of the woodwork. But that was not to be. Harvard University had other plans for the space, and, unlike the situation in 1972, no member of the active physics department faculty was willing to request the continuance.

OTHER EXPERIMENTS – CROSS-SECTIONS FOR RADIONUCLIDE PRODUCTION

Experiments on radionuclide production were planned from the earliest days and a radio chemist, Dr James Meadows, was hired as a research fellow. Over a 5 year period, with a break while the cyclotron was upgraded, he studied various radio nuclides. On his departure for Argonne National Laboratory he was not replaced. Some irradiations were again performed by Dr Robert Schneider when he was in the laboratory as a post doctoral fellow. One such was O (p, 3p) C^{14}. When he left the cyclotron lab for greener pastures at General Ionics he realized that there was a need for more cross section measurements for proton-induced reactions similar to ones that we had made before, particularly to study the long-lived radio nuclides that are now extinct but would have been present in the early solar system. Schneider and Sisterson then measured the cross section and excitation curves for the reaction Al^{27} (p, pn) Al^{26} at HCL. The produced Al^{26} was measured in the Tandem Laboratory at the University of Pennsylvania. When these results were presented at a meeting, Bob Reedy of Los Alamos National Laboratory (now affiliated with the University of New Mexico) pointed out that this and other cross-sections are important for his cosmic ray studies. About that time, also, titanium foils were irradiated for Dr David Fink of the University of Pennsylvania, in order to determine the cross sections for the reactions producing short-lived radio nuclides and the cross sections for Ti(p,x) Ca^{41}.

Schneider, Koehler and Sisterson bombarded a piece of the Bruderheim meteorite with a proton beam of uniform intensity. Ed Fireman of the Smithsonian extracted the carbon using his 'usual' procedure to produce a CO_2 sample, which was sent to the Isotrace Laboratory, University of Toronto for determination of the carbon isotopes using Accelerator Mass Spectroscopy (AMS). The value for the cross section for O (p, 3p) C^{14} so determined was much higher than expected. The earlier measurement at HCL was in error due to a mistake in calculating the proton fluence. Once this error was corrected, the revised cross section was in much better agreement with the historical data and other recent measurements of this cross section made by us at HCL and others.

Cosmic rays interact directly with extraterrestrial materials to produce small quantities of radio nuclides and stable isotopes. In well-documented samples from the lunar surface (rocks and cores) and meteorites a large number of cosmogenic nuclides have been measured. AMS has increased the sensitivity of detection of radionuclieds and has allowed these measurements to make routinely in small samples, a great improvement over the heroic efforts previously required to measure the decay products of these long-lived radio nuclides. Theoretical models have been developed to interpret these measurements so that we can learn about the history of the object under study or the cosmic rays that fell upon it. Most of these models try and account explicitly for the interactions of all cosmic ray particles with all elements commonly found in these extraterrestrial materials. Therefore, good cross section measurements for relevant reactions are needed over the cosmic ray energy range as input to these models. Most cosmic rays are protons (~98% of solar and ~87% of galactic) cosmic rays, so it was thought that the most important cross sections needed as input to the models are those for proton-induced reactions.

Irradiations were made by a collaborative program at three facilities. UC Davis, for energies of 67.5 MeV and below, HCL for energies from ~40 to 160 MeV and TRIUMF (Vancouver) for proton energies of 200, 300, 400 and 500 MeV. The irradiations at Davis were made by colleagues at San Jose State University. With the help of the cyclotron operators, Sisterson made the irradiations at HCL. Sisterson and Vincent made the irradiations at TRIUMF. It turns out that neutrons *are* important since many neutrons are produced in the interactions of galactic cosmic rays, which can penetrate deeply into a body. At depth in an extraterrestrial body such as a meteorite, these secondary neutrons produce most of the cosmogenic nuclides. Neutron induced cross-sections were measured at iThemba LABS (iTL), Somerset West, South Africa using neutron beams at quasi-monoenergetic energies; and using 'white' neutron beams at the Los Alamos Neutron Science Center (LANSCE), Los Alamos. Sisterson and collaborators have been able to show using the cross sections that they measured at high neutron energies for reactions producing Na^{22} (a radionuclide not produced by low energy neutrons) that including only the cross sections for only some of the pertinent reactions leads to calculated production rates that are closer in value to those measured in lunar rocks. Dr Sisterson moved to the NPTC at MGH in 1999 where the program continues.

OTHER EXPERIMENTS – RADIONUCLIDE PRODUCTION FOR MEDICINE AND PHYSICS

It is a curious historical development that one of the most important uses of cyclotrons in the 1930s was the production of radio nuclides, particularly for medicine but also for other physics research. It was noted above that this was, indeed, an important use of the first Harvard cyclotron. Although this were discussed in the funding proposal, it was much less important for the second Harvard cyclotron. There were several reasons. The most important perhaps is that neutron rich radio nuclides are easily produced in a nuclear reactor, either by neutron bombardment of a stable element, or as a product of fission. The cyclotron could produce neutron deficient isotopes, but the intensity was not great and medical needs were served by more intense linear accelerators and special cyclotrons, often with lower energy but higher intensity.

A notable exception was the production of radionuclides for the PhD thesis: "On Gamma Ray Directional Correlations Disturbed By Extra-nuclear Fields" of Dr Gunther Wertheim in 1955. We also note in the publication list that Rh 100 was also produced for the thesis of Dr Hohenemser of Clark University.

OTHER NON CYCLOTRON EXPERIMENTS IN THE CYCLOTRON LABORATORY: THE CAT SCANNER

During 1956–57, while waiting for counts to be recorded on the scalers, Allan Cormack would discuss his ideas for improving X ray radiography. He had already thought about the problem while a lecturer at the University of Cape Town and as an informal consultant to the local hospital. It is stupid, he used to say, to limit the information to a simple picture showing the absorption. This was merely a set of line integrals of the 3 dimensional absorption characteristics. He found an article from 1919 or thereabouts, showing that a set of line integrals, taken from all directions, could be mathematically converted into a spatial distribution. Being an excellent investigator he wanted to show this by experiment. It so happened that the Cyclotron Laboratory had a strong radioactive source as well as a fine colleague in Andy Koehler. So, in the basement of the Cyclotron Laboratory, Allan, along with a student, laboriously measured a set of absorption line-integrals in a phantom. It took a week to convert this into the spatial distribution – a job now done by a computer in a second or less. So it was that the CAT scanner (Computer Assisted Tomography) was conceived. It took Allan's energy and enthusiasm to ad-

Allan Cormack with King Gustav receiving the Nobel Prize for Medicine.

vertise this among the medical community. After a few years he had succeeded and in 1979 he shared the Nobel Prize for Medicine with Sir Godfrey Hounsfield from Electrical and Musical Industries (EMI, UK), who had built the first working example: *"for development of the concept and first experimental models of CT scanning."* We show here our friend Allan receiving the prize.

RIP

In 2002, the University wanted the space occupied by the cyclotron for a large underground parking garage. Although the last treatment was delivered on April 10th 2002 the cyclotron kept going 7 more weeks. The University was not quite ready to begin the process of decommissioning. In the meantime a backlog of radiation damage studies were performed. Mr. Ethan Cascio, one of the many loyal staff members over the years, was in charge of these radiation damage studies in the few last years, and was responsible for the last operation of the cyclotron performing studies for Minneapolis Honeywell. An untimely end came at approximately 9 am on Sunday morning June 2nd 2002. At that time the cyclotron was shut down, by Harvard administrative staff, a day earlier than agreed and switched off for the last time. This was 53 years and 7 hours after the first beam was observed. In summer 2002 the office building was emptied and the look of the control room at last changed drastically.

By October 2002 the office building was dismantled and in November 2002 the shield walls and other material in the cyclotron vault itself were removed. The magnet shims, cut by hand with tin shears by Professors Strauch and Wilson on Christmas Eve 1955 were still in place. The regenerator and beam extraction equipment were the same as those so rapidly installed in the summer of 1956. The magnet, the rigging of which took so much trouble and care to install in 1947, was cut up into small pieces and sold as scrap material. As predicted by HCL staff the radiation levels in the materials were not large and smaller than suggested by state radiation authorities. In summer 2003 the cyclotron vault was removed and the site was turned over to other uses.

But the work lives on. Although Harvard was not the first cyclotron to use protons for radiotherapy it was for many years the most successful, largely because of the close cooperation between the physics department, the cyclotron staff, and the physicians at MGH. Other facilities followed the lead of Harvard. In January 2003 there were 19 other

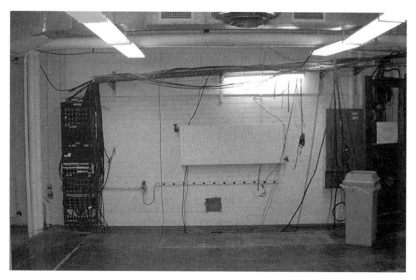

In summer 2002 the office building was emptied and the look of the control room at last changed drastically.

institutions using proton radiotherapy as shown in the following table. In them the Harvard cyclotrons live on.

Acknowledgements

The information in this report has been collected from a variety of sources. The initial collection of photographs of the Harvard cyclotron, and in particular of the medical work, was collected by Professor Janet Sisterson now at Harvard Medical School. These have been scanned by Ms. Yanjun Wang and are available on a CD ROM for those who desire them. Other sources are the website http://oasis.harvard.edu/ html/hua01999frames.html and a paper by Katherine Sopka in 1978, "Physics at Harvard during the past half-century, a brief departmental history, Part I: 1928–1950". Kristen Johnson who worked for a year at the archives after the closure of the cyclotron added a number of documents and lists of publications, as did Dr Bernard Gottschalk. The comments and criticism of many others, especially Professors Norman F. Ramsey and Robert V. Pound, have been important to avoid the innumerable errors in the first draft and somewhat reduced here. I especially thank Kristen Johnson for a critical reading of the final text.

WORLD WIDE CHARGED PARTICLE PATIENT TOTALS

Prepared by Professor Janet Sisterson, January 2003

WHO	WHERE	WHAT	DATE FIRST RX	DATE LAST RX	RECENT PATIENT TOTAL	DATE OF TOTAL
Berkeley 184 inch	CA. USA	p	1954	– 1957	30	
Berkeley	CA. USA	He	1957	– 1992	2,054	June–91
Uppsala	Sweden	p	1957	– 1976	73	
Harvard	MA. USA	p	1961	– 2002	9,116	
Dubna	Russia	p	1967	– 1974	84	
Moscow	Russia	p	1969		3,539	Dec–02
Los Alamos	NM. USA	π	1974	– 1982	230	
St. Petersburg	Russia	p	1975		1,029	June–98
Berkeley	CA. USA	ion	1975	– 1992	433	June–91
Chiba	Japan	p	1979		145	Apr–02
TRIUMF	Canada	π	1979	– 1994	367	Dec–93
PSI (SIN)	Switzerland	–	1980	– 1993	503	
PMRC (1), Tsukuba	Japan	p	1983	– 2000	700	July–00
PSI (72 MeV)	Switzerland	p	1984		3,712	Dec–02
Dubna	Russia	p	1987		154	Dec–02
Uppsala	Sweden	p	1989		311	Jan–02
Clatterbridge	England	p	1989		1,201	Dec–02
Loma Linda	CA. USA	p	1990		7,176	May–02
Louvain-la-Neuve	Belgium	p	1991	– 1993	21	
Nice	France	p	1991		1,951	June–02
Orsay	France	p	1991		2,157	Jan–02
iThemba LABS	South Africa	p	1993		433	Dec–02
MPRI (Indiana)	IN USA	p	1993		34	Dec–99
UCSF – CNL	CA USA	p	1994		448	July–02
HIMAC, Chiba	Japan	C ion	1994		1,187	Feb–02
TRIUMF	Canada	p	1995		77	Dec–02
PSI (200 MeV)	Switzerland	p	1996		99	Dec–01
G.S.I Darmstadt	Germany	C ion	1997		156	Dec–02

WHO	WHERE	WHAT	DATE FIRST RX	DATE LAST RX	RECENT PATIENT TOTAL	DATE OF TOTAL
H. M. I, Berlin	Germany	p	1998		317	Dec–02
NCC, Kashiwa	Japan	p	1998		161	Dec–02
HIBMC, Hyogo	Japan	p	2001		30	Jan–02
PMRC (2), Tsukuba	Japan	p	2001		145	Dec–02
NPTC, MGH	MA USA	p	2001		229	Dec–02
HIBMC, Hyogo	Japan	C ion	2002		30	Dec–02
INFN-LNS, Catania	Italy	p	2002		24	Dec–02
Wakasa Bay	Japan	p	2002		2	June–02
					1,100	pions
					3,860	ions
					33,398	protons
				TOTAL	38,358	all particles

HARVARD CYCLOTRON LABORATORY
— 50ᵗʰ Anniversary

A Symposium Celebrating 50 years of Proton Beams at the Harvard Cyclotron Laboratory (HCL)

Saturday, 5 June 1999

This transcript is taken from a recording and corrected by R. Wilson. Since the slides are not available, reference to them is modified appropriately. In addition comments about slide projectors, microphones, etc have been removed. The result may be inaccurate but many of the interesting stories may be correct.

Unfortunately, Professor Bainbridge, Dr Preston, Professor Purcell, Professor Street, Dr Kjellberg, Dr Paul Cooper, and Dr Steinberg have died. Dr Robert Wilson, Professor Strauch and Dr Sweet are too unwell to attend. We miss them and many others who have worked with the cyclotron or contributed so much to the success of the program.

CHAIRMAN: DR S. JAMES (JIM) ADELSTEIN

The research goals of the cyclotron have been a part of the noble tradition, following the discovery of x-rays by Roentgen and of radio-activity by Bequerel more than a century ago – that has seen the radiation sciences enrolled, both in the exploration of the natural world, and in the care of the sick and the suffering. Certainly, it epitomizes the importance of technology to modem medicine. And, for very parochial purposes, a happy collaboration of the Faculty of Arts and Sciences at this University, with its medical counterpart. And, 50 years are not the end of an era, but the beginning. For out of the phoenix of the Harvard Cyclotron, so to speak, here in Cambridge, is arising on the grounds of the Suffolk County jailhouse the next generation of protons in the service of patients. All of this makes for a joyful occasion. An opportunity to review history, and a chance to acknowledge the efforts of those who have made it happen. The staff of the Cyclotron, both past and present. Those who have worked with it are to be congratulated – especially it's past Director, Andy Koehler, and its present one, Miles Wagner. It's also an occasion to recall the late members of the University, on both sides of the River who have gone before. And I know that Norman Ramsey, and Richard Wilson will recall their efforts and exploits. I should also like to acknowledge my fellow committee members – Michael Goitein, Richard Lahey, Jay Loeffler, Paul Martin, Costas Papalolious, Carl Stohlberg, Herman Suit, and Dick Wilson. Dick, cleverly stepped out of the Chairmanship when he went on sabbatical sixteen years ago and asked me to fill in for him. I'm still filling. But he has continued to be the Cyclotron committee's principal component and advocate. And, of course to Alice Coggeshall who kept us well organized. It's now

our privilege to have the President of the University provide the official welcome. Neil Rudenstine has championed the cause of inter-faculty cooperation among his far-flung elements of his empire. And, I hope he finds in this
one, a source of satisfaction. We, in turn, are grateful for his presence. Mr.
President.

PRESIDENT NEIL RUDENSTINE
Good morning. There's hardly anything worse than introductions – except introductions by Presidents. And, except introductions by Presidents,
early on Saturday morning. But I do say I'm very pleased to be here. Very few
things in the world these days, last as long as 50 years. And, scientific machines, as you know, better than I – are some of the shortest-lasting instruments anywhere. So, we human beings, on the whole, outlive virtually anything made by man any more. And it is something remarkable to be able to
celebrate 50 years of a Cyclotron that's seen, essentially, three different lives.
One, as you probably know, the first one was built in the last 1930's here. It
just began operations, and then it was commandeered by the United States
Government. A group of people arrived, stealthily, in the middle of a night –
disassembled it. Packed it. Re-assembled it at Los Alamos. And used it thereafter in atomic work during the War, leading to the Atomic Bomb. Los Alamos,
at the end, refused to give it back. How you can not give back a Cyclotron, I
don't know. But, it's not just some little piece of equipment that you mislay
somewhere. Anyway, the United States Government was good enough to provide us with funds to build another one – which happened, as you know,
shortly after the War. That Cyclotron had two lives. That makes three altogether. The first ghost-like haunting one of the '30's. The second life was the
present cyclotron first used in Physics. And then, as Jim Adelstein has just
said so imaginatively, in this collaboration between the Medical School and
the Physics Department. I don't need to tell you what important work was
done in Physics here – including, of course, the training of many graduate
students who then went out and did well in other places – as well as those of
you who were here from the beginning. And others not here today. The medical work, I think, has been particularly exciting. It took extraordinary imagination and leadership, back in 1970, when it looked like all Cyclotrons were
collapsing – to be able to not simply revive it, but to keep it going. But then, in
fact, to build a remarkable program with the help of NIH funding, as you
know, mainly on things such as brain tumors, eye malignant problems and so
on. I think it's now, Jim Adelstein assures me, established that at least one or
two of these treatments are the preferred method of treatment. It never would
have happened without the people here. And, others – still in an exploratory
stage – having to do with macular degeneration and so on, are promising.
One never knows. But there's much, much work to be done. And that will
take place at the MGH when the new machine comes on-line. So, all in all, we
can celebrate three lives – thanks to the imagination and skill and early inter-
vention of lots of people. And, of course, when the original Cyclotron was

built – or the second one, I guess, after the War – it was the third largest in the United States. So, in that sense, pioneers on all sides, and pioneers in the medical side, as well. So thank you. I promised everybody no more than five minutes, and 34 seconds – and that's what you're getting. Thank you very much. Thank you for coming. And let me, particularly congratulate Dick Wilson, Professor Ramsey, Professor Adelstein, the various Directors of the lab. And all of those of you, from the Physics Department, to the Medical School, who've worked so well together on what was a truly remarkable insight into what use a Cyclotron might be made of in medical surgery, so to speak. Have a very good conference. I must now leave. I wish you well, and congratulations on the 50th.

RICHARD WILSON

One of the problems of having a Symposium, just before Harvard Commencement is that the President, and Deans are busy with their real work – which is, of course, raising money. I have a couple of preparatory words. The first time I heard of the Harvard Cyclotron as a young, impecunious graduate student, I hitchhiked to my first conference in Cambridge, England. And among other people there, I met Professor Kenneth Bainbridge, who described the plans for the Harvard Cyclotron. And the thing I remember was the neutron beam was going to head straight for the Divinity School, which, of course, it did. Unfortunately, Ken is no longer around to share this occasion with us. The next person that we would have liked to have here with us is Robert Wilson. Although he is still alive, neither he nor his wife, Jane are physically able to come. But Jane sends her regrets. Bob Wilson had a lot to do with the initial work of this cyclotron. The person who is here and was at the beginning was Norman Ramsey. I personally met Norman Ramsey at the American Physical Society Meeting in the end of January 1951. And it was a violation of the usual rule at conferences – when they are very crowded: only short people meet short people, and tall people meet tall people. But, Norman Ramsey had a sufficiently loud voice that I met him anyway. And a little while later he invited me to come to Harvard.

NORMAN RAMSEY

I noted with pleasure that when I got here on time – on scheduled time that there was a nice set-up with the overhead projector, which has momentarily been removed. And, I would like to get it back it. OK. Well, let me say, it is actually a real honor and a pleasure for me to speak at this Celebration of the 50th Anniversary of this very successful accelerator. And that's a very rare thing. In fact, I don't think it's ever occurred before of somebody speaking to celebrate the 50th Anniversary of an accelerator. The half-life of an accelerator is usually about 20 years. And, in fact, that was almost the case here. They had a great party about 20, 25 years ago. To celebrate the close-down of the Harvard Cyclotron. It was supposed to close-down within a week. And, then suddenly, there was a reprieve. NIH had found some funds, and it was kept going a little bit for research purposes. The research was very fruitful for medical purposes, and it's been going ever since. And it's still going, very happily at the present time. And now, I am particularly happy to get the transparencies now – because what I was going to do at the beginning is to say something about the principal contributors. There is a problem that any project of this size. An unfortunate characteristic that the number of contributors is too many. But that is also a fortunate characteristic. Hopefully, there are many young people to contribute. But there's no way of being fair and being pressed, for time I will do so with this first transparency. There really have been many contributors to the project. Although I'm not listing on this in the next transparency by any means all of them. But I am listing some of the principal contributors over the period of the first five years or so. One important name has already been mentioned, which is Richard Wilson. There is also Robert Wilson – Bob Wilson. R.R Wilson. Sometimes I distinguish in these – R Bob, and R. Wilson – that's Richard Wilson. And, RR Wilson, which is Robert Wilson. Or RR Wilson and our R Wilson who is Richard Wilson. And, RR Wilson was the Director from 1946 to '47. Then Ken Bainbridge who was here at Harvard for a long time. But he was particularly served as Director from '47 to '48. I was Director from '49 to '54. And then, the – but the key thing, if you'll note at the beginning – the Directors were only here for about one year. They changed through the first three years. In the first three years there were three successive Directors. Ordinarily, that is a total disaster for a project, I mean, to have that many changes. But, there was a key thing that made it quite feasible which was – there was a very excellent person – originally called a coordinator, and then later Deputy (Associate) Director. Lee Davenport, who served through all this transition period. And he absolutely, I think, really functioned as Director during that time even though others of us had the official title. And, I am very happy to see that he is here today for this celebration. Because I think the success, the initial success of the Cyclotron was largely due to Lee Davenport. And then, there was the Nuclear Physics Committee which included Bainbridge. Again, Robert Wilson, for the year he was here, Curry Street, Ed. Purcell and others. And then, to just quickly go through the staff. Then, there was a large number of people on the staff. The Engineering Design was headed by Bob Grensback. The oscillator was headed by Al Pote. I don't know whether we paid Al Pote, or he was just here as a volunteer. These were the days before TV. He owned one of the larger radio stations in the area. Also, fortunately for us, he owned the largest yacht in Boston Harbor. So it was a great place for the laboratory to have its celebrations and picnics on his yacht. But, he did an excellent job doing and running and building the accelerator, and designing it. And then, the control system required lots of ingenuity. Well, actually, that work was directed by a graduate student – graduate students were interesting at that time. They had been doing technical work during World War II, sometimes as a soldier and sometimes as a civilian. And then, were coming back for their degrees. And, this was true of Leo Lavatelli. So Leo arranged for the basic design of the electronics, and the control system, which turned out to be very successful. And then, particularly, Art Hanson who was heading the machine shop. And then, there was also a lot of important organizational support. One was the Manhattan District. It is a little bit hard to say whether the Manhattan District supported Harvard or not: whether it was the villain or the help. It's the Manhattan District that took the first cyclotron to Los Alamos – for the Federal Government. It provided $200,000 dollars for making up for its having stolen the first Cyclotron, which provided the building that was used. And then, also, very importantly, the Office of Naval Research, who agreed to really negotiate one of their first contracts with Harvard. It was an experimental organization that we had. And people didn't know how big they the laboratories should be, or how they should be run. But that worked out very well. And that (ONR) was actually the predecessor to the National Science Foundation. Well now, having given my names in this fashion, let me now go on to more of an anecdotal history of the sequence of events that led to the Cyclotron, this second Cyclotron.

The first really important sequence would start in 1943, when the Manhattan District stole the first Cyclotron from Harvard. It was one of the more reliably operating Cyclotrons. It was a small one. It was 11 million volts for deuterons which means even less for protons. It was primarily used for deuterons. Primarily used for studying nuclear reactions. And in those days, that was the new subject in physics. And, it went to Los Alamos. And the group that it went to at Los Alamos was headed by a young Princeton Ph.D., Robert Wilson. Robert Rathbun Wilson. Bob Wilson. And, he headed the group – and this served as the one accelerator of that kind at Los Alamos during the War. But then, at the end of the War, he (Bob Wilson) was recruited to come to Harvard. Presumably to spend the rest of his scientific career. And he and Ken Bainbridge got together particularly with Curry Street, to plan for a replacement for the First Harvard Cyclotron. And they did so. They persuaded the Manhattan District, to contribute to pay for what they'd already stolen. And pay a pretty good price. Harvard made a big bargain, because this money went to Harvard. And it probably cost them $50,000 dollars in the first place. And Harvard got $200,000 back. But they also persuaded Harvard to in turn, spend the money on a replacement building. You couldn't get a big accelerator with that sum of money. But you could get a building. And so they agreed – the University, with that money, agreed to pay for it. It didn't cost the University anything. But I mean, they even probably got a little overhead out of it. But, nevertheless, that did provide the basic building. And, with that, then Wilson and Bainbridge, in particular went to the Manhattan District, and to the Office of the Naval Research, which was a new office. And ONR had done very little. It was just at the end of World War II. The Navy realized that they had gotten a great deal of help from physicists during the War, particularly on radar, and things of that kind. And they wanted to, in turn, keep physics alive in the country. And they agreed – and they initially started a very small program for research support. It was not clear that they could support anything as large as a Cyclotron, but they did. And, agreed to give the construction support, and also the building of it. But the budget had to be pretty small. And that led to the selection of what the energy should be for the Cyclotron.

Now, on the chart I state the energy. The choice was that particularly of Wilson and Bainbridge, who probably had some bad luck, and some good luck. The bad luck was that at the time that they made the choice, pi mesons had not been discovered. And there was evidence from cosmic Rays that no Accelerator could ever produce neutrons, and therefore, there was no particularly obvious threshold of energy. And there was also no particularly obvious threshold of how much money the Office of Naval Research could put for a machine. So they chose a bit conservatively. And chose an energy of about 130 MeV. Unfortunately, after this design decision was made, the pion was discovered and it showed that had the energy had been more, you could have produced pions. On the other hand, it was still a very useful machine for studying nuclear forces – which is what it had been originally designed for. And, so – that was the bad luck. If they had known about pions earlier, they could have persuaded ONR to have gone a little bit higher in energy (and money). The second bit of the good luck that compensated – it turned out that was just the right energy for proton therapy. Well now, proton therapy wasn't known at the time. So, that was not in their plans. But that was, I guess, in the long run, has been a fully compensating bit of good luck. A little worse for the physicist but certainly better for the medical people.

Well then, Robert Wilson was appointed Director of the Laboratory. And, he realized that most of the talent and knowledge about designing accelerators existed at Berkeley. So he really spent most of that first year, and, as it turned out his only year at Harvard, in Berkeley. Designing, using their facilities and their advice on designing, the Harvard Cyclotron, which designs he was then going to bring back to Harvard and get it built. While he was there he was also is a pretty active physicist. He did a very important experiment, from the first experiments at comparable energies, on proton-proton scattering. And then, he wrote one very important paper, which was the first paper suggesting proton therapy. And in the paper he noted that there is the thing known as a Bragg curve – when a high energy proton goes through any material such as water or tissue. As it gets near the end of its track, the proton is going slower and slower. And since, it's slower, it spends more time in the presence of each atom. Therefore, the electrical field acts a little longer, and gives a bigger impulse to the electron. And, therefore, the ionization is greater at the end of its track, than it was at the beginning. Well, since the main problem one has on using radiation therapy is trying to damage the cancerous tissue as much as you can with damaging the normal tissue as little as you can, he recognized that if you could bombard a patient from different directions, but always try to make it such that the protons were near the end of their range, where they were in the cancerous region, this would significantly increase the damage done to normal tissue. It would increase the damage done to the cancer, and diminish the damage done to other tissue. So he wrote this up in a paper in 1946. And, then came back to start his work on the building of it here, and finishing the design. But then, in 1947, one year later he was lured away to Cornell.

And, then that same year, I was lured from Columbia to Harvard a fact which did not have anything to do with the cyclotron. I was supposed to be setting up molecular beam research. Study on radio frequency spectroscopy, which we had then been doing very well at Columbia. Ken Bainbridge actually succeeded Bob Wilson as Director of the Lab and as Chairman of the Nuclear Physics Committee. And then, the following year – after I was here a year – I guess they felt that now I had been caught properly at Harvard. And then, they advised very strongly that I should become Director of HCL. Bainbridge was anxious to get on with his mass spectrometry. So I agreed to be Director. This particularly became feasible because of the excellence of Lee Davenport who was the so-called Coordinator of the project. And he provided continuity for the work that was done there. And I think that the two of us had a very happy time in running that laboratory, from that point on, for the next few years in the next phase of construction. The design period really gone from '46 to about '48. 'But, some design changes continued on to '49. But on the other hand, construction began in some parts in '47. Some contracts were let out before I was involved but the principal activities on construction occurred after I was Director. And one of my most valuable contributions was recognizing that the Cyclotron was a very well-designed machine. And therefore, I shouldn't make many changes. I should just see the design through. We didn't have a lot of changes to make. And, secondly, it was a very good design. I think for this, Bob Wilson, deserves much credit. I think we had the good luck – it wasn't merely that he was so tall in that connection. I think he is probably the world's most talented accelerator designer. He, in fact, became the Director at Fermi Lab – which is still the world's highest energy accelerator ever built. In fact, there at Fermilab he and I had our second collaboration. And in that case, I was President of URA. And I had to hire the Director, and work together. But he was also the Director, and his ingenuity and design worked out very well in that highly successful accelerator. But, in any case, his design here was clearly very good. So, we basically followed that. And as I say, I make a useful contribution of not making too many changes. Then, we did concentrate on trying to do our best for achieving reliability and the design was such that it could be easy maintenance. And, I think Alice Coggeshall warned me that I had better be careful about that statement, because the maintenance people in the lab will say, "boy! it's not that easy to maintain!". Well, it's not easy to maintain. No accelerator is easy to maintain. But, compared to almost all other accelerators, this one is relatively easy. It does run quite well for accelerators of this energy.

The first experiment was done during the construction phase – before it was finished. It was sort of a joke almost. Namely Purcell had just invented NMR – the two of us were closely working in that general field of resonance. And, we began speculating. Could, by any chance, the proton's spin orientation be related to memory storage? Because you've got a tremendous amount of storage if you have every proton in your brain, determining that lots of bits can be stored that way. And we didn't think it was likely. But you know, it might be. So, we devised the following experiment. This was in the days before OSHA. And before you had to get approval for experiments involving humans. Ed made up a coil and an oscillator. He brought an oscillator over and I turned up the magnet. I knew how to run the Cyclotron magnet then. So, one Sunday evening afternoon – when I turned up the magnet, we tossed a coin to see who went in first. Ed went in first. And put a coil around his head. And then I cranked it through the proton resonance. We weren't sure what was going to happen. Maybe he'd say "ouch". The worst would be maybe to wipe out his memory. (laughter) Then, he'd start all over. And, well – nothing happened. (laughter) So, then, it was my turn. Nothing happened with me. So, we sheepishly folded up the apparatus. Obviously we didn't publish this experiment. (laughter) But except somehow that schpiel got spread around. And there are several books on the history of MRI – Magnetic Resonance Imaging. which quote this as the first Magnetic Resonance Experiment on humans. Well, it's not our proudest experiment. But, nevertheless, we did do it.

But to return to the construction. Primarily the design was a really very good design. That doesn't mean there were no construction problems. I mean there were construction problems. This was still fairly shortly after World War II. And, there was a shortage of everything still. And, deliveries were badly delayed. But, somehow, we got around them. And went really surprisingly rapidly. And on that, Lee Davenport was an expert at expediting what we were doing. And we also had some problems. Although the design was excellent, there were some problems – one of which pertained to leaks – vacuum leaks. There, Berkeley accelerator experts – the experts at Berkeley – just at the time that this accelerator was being designed persuaded us to buy a new stainless steel that was non-magnetic that had just been developed. They thought that this was going to be the greatest thing ever. They'd never used it, but they thought it would be good. So, they convinced Bob Wilson, and Bob Grensback, that's the material to use here. So we got that ordered and cut. And then it turned out it was a great material. For it had the magnetic properties desired. But nobody could weld it successfully. It was a real problem. We had arranged for the welding of the vacuum tank to be done by the Navy Yard which had expert welders. I mean they'd been welding all sorts of ships during World War II but they couldn't weld this very well. Neither could anybody else in Massachussets. Nor could anybody in Berkeley. They never had used this. But, we'd finally forced it through – but, with lots of worries, and unfortunately, with a considerable effort spent on leak hunting. I remember, in fact, one time when we were having this misery tracking down the leaks, and getting it re-welded in places, that in those days, the President of the University had a formal reception for the members of the faculty. You know, black tie, tuxedo, and the women in long dresses. And while we were at the reception, Lee and I were talking about possible other places that could leak and we had a good idea. So, we said we'll go back after this reception, and hunt for leaks, which we did. We were in our tuxedos, and leak-hunting. And I really felt very badly as we were doing this, that there was, at that time, no camera in the laboratory. Because it would have been marvelous to have had a photograph of Lee, and myself, in tuxedo leak-hunting, and then we could show the transparencies – and say, this is how we leak-hunt at Harvard. (laughter) But, unfortunately, that picture exists in my mind, but not otherwise. Well, another problem that we've had – particularly on leak-hunting. It turned out – one very – one of the pumps was very inaccessible. Very hard to get to, to do the leak testing on. And, that had been tested by the manufacturer. Guaranteed to be absolutely tight. So, we sort of hoped. But, we still kept having a rather persistent leak, which we thought was due to the vacuum chamber walls, but it seemed not to be there. And, well finally, it turned out that after the pump had been made by the manufacturer, and tested fully. But then wanted to put the label on, naming the manufacturer. And, in the process, they drilled a hole through the side. And, although it got pretty well plugged by the screw, it leaked just a little amount and a small leak was hard to find. But at the same time, it was critical. One of the final problems happened after we got everything going. We got the magnet going. We got the oscillator oscillating well. We had the vacuum system going well. But we couldn't make it work together. When we turned on the magnetic field, the oscillator failed. Turned off the magnetic field, the oscillator was fine. And, we all stood around for a while. It finally turned out that someone had left a screwdriver in there, on top of the magnet pole. When the magnet was off, it laid flat on the bottom, and caused no problem. When the magnet was turned on, it rose up on end and shorted the oscillator out. So we got that problem overcome. That is typical of the sort of problems that you have. But finally, we got our first beam on June 3rd of 1949. And, really got it operating.

This really is, very accurately, the 50th anniversary. It always takes a big interval of time between the first beam, and reliable operation. Sometimes it is a year or so. In this case, it was pretty short, but it was still a time. And I'd say that I noted one thing in one of our data books; a notation that we had reliable operation, beginning September 30th, 1949. 'That' was a couple of months later. Now, of course, after the war, I'm sure some Presidents usually say there's still no reliable operation. No one ever considers an accelerator is totally reliable if he's a user. But, nevertheless, from the point of view of those who build it, it was going really very reliable bt September. In conclusion, I just want to say about what happened after the beginning of operation. In the first place, we fairly early concentrated on this desire for higher energy than had been designed. We did manage to squeak the energy up from 130 to 160 MeV and maybe we could go up to 180 or 190 if we pushed very hard, but it would not be very reliably there. So, we didn't. 160 was about the highest. Which was still below the pion production threshold. But was very good from the point of view of sundry and fundamental physics.

Subsequent speakers are scheduled to talk about the physics results of the experiments. So I will say nothing about them. I'll only make comments on about three different things, pertaining to the operation of the machine, or the things that didn't pertain to physics results. One of features – that worked out very successfully – was that the machine was designed to be capable of being operated by the students who did research there. We had a day-time crew to run the machine, and help train students. But then, if they wanted to run all night, as you usually had to do, they could run the machine themselves. This worked out very well. The machine, I think – in that sense – was relatively easy for them to learn, and operate. We had one incident that was close to a catastrophe. I think I almost lost a graduate student to a heart attack on that. And this was a very good student. Uli Kruze. And he had gone through his training period with the professional crew. He was still in his first when he was in charge in the evening. And another graduate student was with him for safety reasons. But, other than that – he was on his own. He went through his check list. And he was already to go. And he pressed the "On" button. And, immediately, all the lights went down in the building. The relays began to clash and in and out, making a huge noise. He didn't know what it was all about. He looked out the window. And the lights in the surrounding Cambridge were going up and down. I was home in Belmont when I knew he was having his first run. And our lights were going up and down. So, I rushed in to the lab to see what was happening. Eventually, it turned out that it had nothing to do with the cyclotron. It was pure coincidence. It had absolutely nothing to do with his pressing the "On" button. Somewhere in Western Massachusetts, they were doing some work on one of the main power plants, and with one of these air hammers, they were drilling through two places, and they short-circuited the main bus bars for the electricity grid in New England. All of Massachusetts was off. New England was off for about a day while they were rectifying the problem. But it was pretty panicking for years. And I must admit that I was a little bit ill at ease too. I will also mention our contribution to Alan Cormack's Nobel Prize from our lab. The first prize that anyone from the cyclotron had been awarded. Namely, Alan Cormack was a post-doctoral research fellow, working with Dick Wilson and myself. We were paying him from our funds. And, we knew he was boondoggling some on the evenings. He was doing calculations on his own. But that's OK. We were very nice. We let him do it. Well, it turned out that what he was doing was doing the theory for interpreting CAT-Scans. And for this, he received the Nobel prize. And I like to think that we are partly responsible. If we had a well-disciplined lab, we would have his boondoggling. And then, he couldn't have done it. So we gave a sort of negative contribution to his Nobel Prize. But I think he was delighted when he got it.

On the whole we had good relations with the biologists who were worried whether our radiation was going to upset their instruments. So we decided we are going to do a super careful check. We got one of the technicians with a cart with a very sensitive radiation detector. And, a walkie-talkie. He walked around in different locations. And, the levels were down very low, because it was even further way than we usually looked. But then, suddenly, as he got to the entrance to the Biology labs, he got a great big signal. He got on his walkie-talkie: "shut down the Cyclotron right away! We've got to shut down." So we shut down the Cyclotron. He still got high readings. Well, then it turned out that the granite in the very attractive entry-way (between the herbarium and the museum) was quite radio-active, and produced radiation levels way above the tolerance that we would allow for our neighborhood. From that time on, we've had no subsequent complaints about radiation from the biologists. And they are all the complaints we got from anyone as a matter of fact. And now, I'd like to conclude with just a couple of quick pictures of how the Cyclotron looked in its earlier days.

Here is a photograph of a view of the machine with myself and Leo Lavatelli. He was a graduate student. A pretty senior graduate student, because he had been doing technical work during World War II. And he had a job – and one of the useful things that the Cyclotron did in its early phases of being built was that at that time, there was no NSF. There were no NSF fellowships. And there was a problem of support for students. Some of them had support, some of them did not. But one of the key sources of support was actually came from the hiring graduate students to do much of the building of the Accelerator. They got experience out of it. And, in particular, they also got some financial report. And I think, this, I think I will close. It is time to move on to some of the results with it. I think it's been a very pleasantly successful operation. I know that Lee Davenport and I were talking just earlier. I mean it's had a much longer life than we expected. The normal life of an accelerator I would say is 20 years, or less. By that time, it's usually out of date. And no longer interesting. And in this case, well in a certain sense, that was true of this accelerator for its first career – namely as a physics instrument. There isn't much with it. But, it then had this great new career in medical treatment – particularly in proton therapy. And I am just delighted to see the way that it is going. And, we know that about 30 years ago, I attended a marvelous party to celebrate the close-down of the Harvard Cyclotron, which was supposed to occur a week later. It was revived. And, it's still running. And it seems to be a cat of many lives. And so, I wish to express great pleasure to all. And, particularly, the people who supported it at that time. Thank you.

RICHARD WILSON

Now I am going to talk about the upgrade of the Cyclotron. And that just synchronizes with the time I came here, which was 44 years ago. In 1955, why did we want to change the Cyclotron? It was working well. It was doing some nice experiments. Well, there were two physics reasons. Firstly, like all machines, you want to get the beam *out*, get a nice external beam. To only have a beam going round in circles inside the magnet was not a good idea. You want an external beam. And, secondly, we wanted to increase the energy. And the primary reason at that time, was to get a high polarization of the beam. And just to remind you, it turned out, if you scatter protons off a spin-zero target, you can get nearly 100 percent polarization. And this is, by the way, the figure that we'll mention later which was taken here. This particular spin-zero target was helium. If you have 150 MeV, you can polarization to a certain angle, 95 percent. If you go down to 60 MeV, the polarization is only 15 percent. So, we wanted to get up to that higher energy with the high polarization. And it was clear already from experiments of Chuck Oxley at Rochester and other places that this was important. So when I came, this was just the proposal everyone was talking about. And so, the first thing we asked was how do we increase the energy? When it turned out it had been running at 90 MeV wheareas it had been designed for 130, but wasn't actually running at 130. We have to improve the field at large radius. Karl Strauch and I particularly coped with that. We had to shim the magnet. And the thing that I remember was that Karl Strauch and I were there until 1 :00 p.m. on Christmas Eve with tin shears, cutting shims. We were putting them inside the magnet. And they are still sitting there, by the way. But for us the thing important thing was that for some strange reason, our wives forgave us from being away from the family on Christmas Eve. But that happened. Then we had to modify the Oscillator to make sure it covered a wider frequency range. And the responsible people there were Andy Koehler, who's still hereof course, and Paul Cooper who unfortunately died a few years ago. We always think of Paul as the last of the Fenimore Coopers. And he was a graduate student at the time. Andy Koehler came to the Cyclotron, I think, one year or so before I came. And so, this was one of the first things that he was involved with. So, then came the beam extraction. And what's the problem with beam extraction? Or course, you have to be able to get the beam out. The problem came that when you have the beam in the Cyclotrons in those days, we used to have very large radial oscillations around an equilibrium orbit. But you couldn't have very larger vertical oscillations, or you'd hit the poles. And when you got a particular value of "n" which is a particular parameter of the magnet field gradient [n = (r/B) dB/dr]. When n is equal to 0.2 there was a coupling of vertical and radial oscillations. And that inevitably happens in an accelerator. And so what happens is that as soon as the particles accelerate and get out to radius where n equals 0.2, all the energy in the radial oscillations goes into the vertical oscillations and you lose the beam because of the relatively small vertical gap. The first solution was by two bright people, Jim Tuck, an Englishman who went to Los Alamos and Lee Teng, of Chicago. They thought of the idea of a regenerator and a peeler. They decided that to have coherent oscillations, you must interact with the particles as they are going around You first push them in, and then you pull them out and you repeat the process. For this you have a regenerator, and then the peeler. The peeler to pull them out, and the regenerator to push them in. And, they tried that out in Chicago. And it was never used very much, because Le Couteur in Liverpool came up with a better idea. He said, that you don't really have to peel at the one location. You can be pulling the protons out all the time because the field falls off anyway. All you have to do is be pushing them in coherently at one location. Effectively, that makes the average beam parameter "n" equal to 0. The average parameter to zero. And this was first tried at Liverpool, and then at Chicago. And then, we were the next people who used it. In spring 1955, before I came here, I went up to Liverpool, especially to see that beam extraction and talk with Le Couteur, and we had a design. Well, we finished this work, and we wrote about it in a paper Gerry Calome et al. (see reference list) One of the authors, Paul Cooper is not here unfortunately. He died of a heart attack in an island off Australia. I don't know where Engelsberg is, but George Gerstein apologizes for not being able to make it. He's in Europe at the moment. Andy, of course is here. Arthur Kuckes is here. I don't think Jim Meadows managed to get here. Karl Strauch is unwell and has not been able to get here. And myself. So those were the people. I would like to have some photographs of the whole regenerator – but that is now inside the Cyclotron. (The photographs in the main text were taken after the shut down; see page 21).

We had a number of ideas which were unique to this cyclotron. We decided that we'd have more than one place to get the beam out. The one which everybody now uses. We'd get the external beam out. And the other was a partial extraction of the beam. To extract it to just hit a target. And this was unique for us, I think. But I'll explain why we did it from the physics point of view. In this schematic (from the paper) we show what the regenerator actually is a couple of pieces of cobalt steel concentrating the field in this location, as the radius of about 42 inches. And then, in order compensate in order to make sure that the field is the right sort of shape, we put all sorts of shims out here. And, then we had to match Le Couteur's careful calculation with what we can actually achieve. And we actually achieved pretty well what Le Couteur recommended. Then, of course, how do you measure the field? Again, it was a graduate student that designed the equipment, Arthur Kuckes. One of the best experimenters that we've ever had. He introduced me to a DC amplifiers called flip-chips I think they were called. Maybe they were Philbrick units. And he found them, and bought them. They were to integrate the induced EMF. Normally to measure a magnetic field you have a coil which is flipped over and the EMF is integrated. We were laboriously measuring the field by taking the coil at each location on the orbit, flipping it over and integrating the EMF. Arthur said, let's not do this. Let's just get the core and let's rotate it around the center of the magnet along a particle orbit.. And this he did. And automatically as we rotated the coil, we measured measure exactly what we wanted to know the line integral of the field. So those were Arthur Kuckes' contributions. And I think he took only a very short time. I think it was about 6:00 o'clock at night is when he and I were discussing this. And by the time I came in the next morning, Arthur Kuckes had made the devices. It was certainly one of the things which was positive was the speed of doing things was rather greater in those days than now. Then, we came to a day in April 1956, the first external beam. We put a little probe... an ionization chamber on a probe we pushed in on the machine. And the idea was that if the beam was coming out this far and the beam had jumped over the gap, where the wall of the ionization chamber was, we'd get some ionization. So, we accelerated the beam in to a large radius, and we got to where the beam was supposed to be striking the edge here. We saw no ionization. But then we think we did not completely understand it. And, it wasn't until we pulled the probe a little further out that the beam oscillated with a large enough amplitude that it was no longer striking the edge. We suddenly realized that we were extracting the beam. Although we hadn't really quite expected it. It was about 1 am. It was all written up in the lab notebook which I could not find again last week. So, of course, we celebrated, and Paul Cooper went back home. We called up Norman and we woke him up, and he came in to help celebrate. With a bottle of 1837 Madeira, which Paul Cooper came back and provided. The actual label on that bottle was put in the notebook. But, the notebook seems to have vanished. Of course, the bottle of Madeira vanished much quicker.

We realized that the distribution in time of the beam corresponds to the distribution in energy of the equilibrium orbit. The time distribution of the beam, without any beam extraction, was long. But when you start getting the beam oscillating radially, the particles hitting the target are not just those with the equilibrium radius equal to the target radius but some with a smaller equilibrium radius because they happen to have large oscillations. Associated with the spread in equilibrium radii, the width of the time distribution was large. If we force the beam into a coherent oscillation the time distribution comes down, and so does the energy distribution. We could install a clipper to try to reduce the spread of oscillation amplitudes and then the time distribution becomes much tighter. And that corresponded to an energy width of about an MeV. For some reason, I've never been able to figure out, we were never able to actually produce that energy distribution in an experiment. Well, at least I wasn't. I think Bernie may have been able to.

We then began the nuclear experiments. And I'll just share just one of them, before getting onto introducing the other speakers. One of them was presented at conferences by Allan Cormack. Norman mentioned Allan Cormack. This is particularly scattering the protons by helium. There is a story about that experiment. We were running around the clock. One Saturday night Allan Cormack was on night shift. I came in at 8:00 o'clock to relieve Allan, on Sunday morning, and Norman Ramsey was coming in shortly afterwards, about 9, I think. But, Allan had lost the beam. And my contribution was to notice that the magnet current had gone too high. He lost the beam because the magnet regulator had suddenly gone to pieces. We had a mechanical feedback system from a sensor to an electric (Selsyn) motor to drive the Variac in the power supply. And that mechanical link was a chain link with some limit switches. And the drive had gone beyond this limit switch and broken the chain. And we had to fix it. I figured out what was wrong. Then, came in Norman Ramsey who figured out what to do. We replaced this chain link by an O-ring, which goes over a pulley which slips when it gets to the end of the drive, rather than breaking anything. Then Allan Cormack went to the machine shop, and made the pulley. So by 10:00 o'clock in the morning, we were back in action. So, those were the three of us. And of course, you can see that Norman and Allan they did the work – which is of course the reason why those are the people who got the Nobel prize.

We had a fairly simple way of coping with hydrogen targets in those days. We had no OSHA. And actually I thought it was complicated. For when I first used a liquid hydrogen target at Oxford, I just went to the hydrogen liquefier, got some liquid and opened up the valve. I filled up a glass Dewar with liquid hydrogen. Put it in the back of my car. Drove it 50 miles out to the Harwell cyclotron, poured it into the target, and we started running the experiment within a few minutes. I was a little careful driving the car out to Harwell. I went slowly over bumps, and put a little tube out of the open windows to make sure the hydrogen did not accumulate in the car. But, the cyclotron target was more complicated. The liquid hydrogen was contained in a tube made of mylar. I'm not sure safety people would let us do that now. We found out that mylar worked at low temperatures. If the target wall had been aluminum, it would break. We transferred the liquid hydrogen from the large 40 liter Dewar flask to the target. I can't remember whom, but some graduate student brought the first liquid hydrogen from Brookhaven in a rental pick-up truck. He had trouble when he crossed the bridge over Long Island Sound. We did have proper labels. Liquid Hydrogen, caution, flammable, etc., etc. But when he got to the toll house the tollkeeper said, "You are not allowed to bring that stuff here, but, go on quick." Those were the days before OSHA rules. I've just been trained as a target operator at CEBAF. And it takes about two days of training how to cope with the special complicated refrigerator and target. Which has far less hydrogen in it than we had; and it is cooled by liquid helium. And it has an enormous number of complications. It has three computers working to keep the temperature constant and so on. Life has gotten a little more complicated.

I will tell one other story that I will tell. And, just after we'd started operating again, George Gerstein was running the machine. He just couldn't get the cyclotron working. So, I came in at midnight to see if I could help him. And we struggled for about 2 hours. I had installed a window on the edge of the vacuum tank so that I could look in, and have a look at the ion source. I always like seeing things with my eyes, or handling with my hands. And so, we switched the magnet and oscillator on. I could see the bright light of the ion discharge. The magnet was on. We had a little trouble keeping the oscillator on, but with a bit of effort we got it on, with a higher DC voltage than usual. But, we could get no beam at all. Everything seemed to be otherwise all right. So as we were looking at the ion source, I realized that I had some magnetic keys in my pocket which I should have left outside. These keys were horizontal. Why? They should follow the magnetic field. The field shouldn't be horizontal, it should be vertical. Someone, just that day, had disconnected the coils and they had been misconnected. The top coils were inverse to the bottom coils and the thing was connected as a quadrupole, instead of an ordinary dipole magnet. And so, the cyclotron was, of course, not working. This was found at 2:00 in the morning.

So with that, I will call on Jacques Lefrançois who is a French Canadian living in Paris. And we have had two other Canadians, who were born in other English -speaking parts of Canada. So we have had a nice mixture here.

JACQUES LEFRANÇOIS

Right away, I live in Paris, because Dick send me there for two years. I retired 8 years ago. So, I prepared a few transparencies on the program of the Harvard Cyclotron. And, of course, I've only seen a short part. But I tried to collect either my memories, or articles on what happened. And I will tell you a bit about the motivation of that physics' program. The why, and the what of the program. And then, also I will say a few words, which are more personal, on the life of a Ph.D. around 1960 which is how we did these experiments. Now, motivation. Now, I have the excerpts from the first team. So, these were very serious people. You've got two Nobel prizes in there, so you have to be-lieve what they say. And, luckily they got the Nobel Prize not for the pro-grams. We only worked with the Cyclotron, so we didn't get invited to Swe-den with them. So I read that, in the investigation of nuclear force, the nucleon-nucleon interaction bears a role of primary importance, they said. Its relative simplicity rendered it more susceptible to analysis than interaction involving heavier nuclei. And they went on to say that maybe there's a hope that by understanding the nucleon-nucleon interaction, you could calculate and pre-dict what would happen on complex nuclei. And this turned out to be partly true, the second part. But, the first part not as well. Now, the other point is that this is what they said. But, my memory of a young graduate student was much more hopeful. The idea is that we wanted to understand where the nuclear forces were. And what was clear to us was to understand, or to me was that we'd have to study elementary particles in interaction. Because if you want to understand something, you should put yourself in the most elemen-tary situation to understand what happened. That was the understanding of protons and neutrons. Now, in 1960, when I arrived there, they were very careful not to say that. But, of course, young graduate students say, what can be more elementary than a nucleon? This just shows how naive we can be. Because we now know better. Now the proton, the neutron... these are now known to be very complicated objects. They are made of quarks, gluons and virtual core quantum type block pairs. And, in the 1950s we understood a bit of the complication, because in the 1960 language, we knew that there was a pion cloud around the proton. Around the core of the proton. And, actually, we also knew that we could calculate part of the nuclear force by taking in account the fact that this pion cloud existed and was being exchanged be-tween the two nucleons. But this was only a small part of the whole story. And, actually, the whole scattering was much more complicated. And it was complicated because the proton is complicated. And there was no hope of getting any fundamental interaction by doing proton-proton scattering at this low energy. It's like trying to understand the electric force by doing the scattering of two big neutral molecules, you would have no chance.

Now, nevertheless, I think we all learned quite a bit from doing the physics there. And what I would think I've learned is that if some measure is believed to be important, then it should be done well. And I will try to explain what "well" meant. Since nucleons have spin, scattering will depend on the spin orientation. There will be a spin-spin force. And, a spin-orbit force. And a complete interaction should measure those effects. Now, in a very nice article and review of nuclear science by Wolfenstein characterized completely this interaction. He characterized the scattering by five complex functions of energy of angle. He put in another condition, because the reactions can only be elastic, because we are below pion threshold. This means that the mathematical functions must be real. That was one good thing of being below the pion threshold. And then, you decide that you have five independent numbers at each angle of energy. And, if you have five unknowns, you need to have five measurements. So, we needed at least five different experiments to obtain these five independent numbers. So, doing the program well meant doing five, or more than five experiments. And, these experiments are written here. Cross-section. Polarization. Depolarization, Rotation, A'. These had to be measured for proton-proton scattering and then for proton-neutron, or neutron-proton scattering. And I'll explain a bit what these things are about.

At this energy, when you send proton on a carbon target, and look at the right angle, around 13 degrees or so, 14-degrees, you get a very high polarization of the scattered proton. If you send a polarized beam, it will be only be scattered to one side. If the spin is up, scattered to the left, if the spin is down, scattered to the right. Then by using another carbon target as an analyzer of polarization by the left-right asymmetry, you can measure the polarization of your proton. Conversely, if you send an unpolarized beam onto the carbon target you get polarization. Since the one which have spin up are scattered to the left, and the ones which have spin down, are scattered to the right. If you look at the beam which is scattered on one side, and one direction, you get a polarized beam. So, this is how the program was done. The proton beam was sent into a carbon target. Very close to the output of the cyclotron, actually. And, you extract the beam which was polarized. Send it onto a hydrogen target. And then, you measure a cross-section by looking at the number of scattered particles as a function of the scattering angle. And you measure the polarization, which is how hydrogen scatters to the right or to the left to polarize the proton which comes in. So, hydrogen was used as an analyzer there to get the polarization function of hydrogen.

Then you do the more complicated things. You take the polarized beam and allow it to scatter on a hydrogen target. And look at how much polarization there is left after scattering. You measure this by scattering again on carbon, and looking both at the right and the left. This is called "D" for depolarization: if D = 1 there is no depolarization. This was the first of the triple-scattering experiments. You scatter first on carbon, secondly on hydrogen, then on carbon to analyze again. The second triple scattering experiment was done by putting a magnet with a field along the trajectory of the particles. The spin of the proton, which was up, was rotated by the field. When we calculated that it had rotated by 90 degrees, then we scattered again on hydrogen. You look at how much spin you have left, by doing your scattering on carbon up-down instead of left and right. This measurement was called R for Rotation. And then, there is an experiment with another spin orientation. You do that by having a magnet with the field perpendicular to the trajectory. The spin rotates again at this plane until now it goes along the direction of the particle. You then look at how much spin there is perpendicular to the particle afterwards. And to do this you scatter again up and down to measure the parameter A. You want to look at the component of spin which is longitudinal after the scattering. before the magnet. We installed a magnet to rotate this longtidudinal component into the transverse direction when it could be measured by comparing the scattering up and down. This is parameter R'. And the next one, which wasn't done, which is A' would be to scatter a beam with longtitudinal polarization before the hydrogen target, and measure the longitudinal component afterwards. The need for two solenoid magnets is the reason why it wasn't done, and it was not needed, because this last one, A' is a simple function of R', A and R. In principle, one could also do other more complicated experiments such as scattering a polarized beam on a polarized target. But we had done six measurements we only needed five. That was a complete set of experiments at all angles. We don't really need "all angles", because we are great believers in the fact that reality is continuous. So the curves must be rather smoothly varying. And how smooth depends on how many partial waves you have in the amplitude. How you decomposed the amplitude into "partial waves."

And, at a low energy you find that you have a single wave and the cross-section is constant. It could vary as $\cos\theta$ or as $\cos^2\theta$. At higher energy there could be higher waves. A way to formalize this, is to do a "phase-shift analysis." This was done by a team at Yale, which analyzed other experiments, and made predictions of what we would measure. They parameterized our ignorance by several phase-shift analyses which had several phase parameters at each energy. Then they predicted the experimental numbers at each energy and angle. But, when you have a certain number of unknowns, and, a limited number of experiments, there are sometimes ambiguities. Yale had various solutions. Which they numbered YLE0, YLE3, YLE3M and so on. Now, this was a parameterization of about 15 different numbers at each energy.

There was also a hope of finding a potential to describe the interaction. Now, a potential would have been very nice, because it would have been trying to understand the real force. Once, if you find the electric force, and you say that you have a potential function $V(r)$, the force varies as $1/r^2$. And we understand the electric force. So, if we could find a potential, that may be we were getting closer to understanding the nuclear force. But, we had no luck. It was not possible to describe this complicated situation with a simple potential. We knew that the potential was attractive, because nucleons stick together to make nuclei. But, the nucleons don't coalesce. So it must be also slightly repulsive. So there's a repulsive core, surrounded by pions which attract. The potential was different for every spin state. So, in all, theorists had a potential which had 14 parameters, and even then did not the data very, very well. So, again, retrospectively, there was no hope of simple parameterization of complicated physics.

Now, let me say a word on the problem of the neutron-proton experiments. You have two choices. If you want to do neutron-proton physics, either you send neutron beam on a proton target, or a proton beam on a neutron target. Now, neutron beams are less intense, because you have to produce them with your proton beam, so that is decreasing the intensity of the beam. You cannot bend them, ionization them in energy. You can not force them with quadrupoles, so the beam is not very well defined. So, there were some experiments done, but they were very difficult experiments, because of this. Now, the other way would be much easier to send beams on neutron targets. But of course, pure neutron targets do not exist. And the idea was to pretend that the neutron in a deuteron in a liquid targets is free. This target was designed by Hoffman, for his Ph.D. thesis. This deuterium target which was a bit more complicated than for the liquid hydrogen that Dick discussed. And, you pretend that the deuteron is actually a free proton, and neutron. And you make some correction for the fact that the proton and neutron are bound, and the deuteron slightly bound. This means that the neutrons have a Fermi momentum going around. So there's a smearing of the nucleon-nucleon kinetics. one particle hits at a target which is moving at random in various direction. But you can correct for that by doing some kinematic calculations using the Fermi momentum for which wrote a program on the computer at Harvard. This was a very, very slow computer. One thing you can notice, is that when you hit a proton with your incoming proton beam, then there will be your proton which is scattered, and there will be a proton recoil. And, when you hit the neutron, there will be a neutron recoil. So, you can separate event by event, whether you've hit the proton, or neutron. And then, looking at the case where you have the proton report, and comparing it to the case of a pure hydrogen target, you can check whether your assumption was valid. And, so both experiments were done, and p –> np experiments were performed. And there were also some elastic pd experiments. This may fall in-between what Bernie is going to talk about – the nuclear physics. And what I'm discussing – proton, elastic scattering on deuteron – I would say was really used to determine the phase-shift analysis, or things of that sort. Because the physics is a bit more complicated. But, it was a nice test of the first idea that you can calculate nuclear interaction if you know the proton-proton, and the proton-neutron interaction.

There were predecessors for the nucleon nucleon program. Owen Chamberlain, and his colleagues at Berkeley, at the Berkeley Cyclotron, at 310 MeV had done a series of experiments like those. But the energy was above the pion production threshold, and this complicated the analysis quite a bit. And, then we had competitors at Harwell, U.K., and Orsay in France. But they were doing only part of the program. They didn't get around to do the full set of measurements. And then, there was Rochester, which had done one experiment before us. But then, they had a more complete program after Ed Thorndike, who had started with Harvard Cyclotron, went there. I have tried to make a list of everything that was measured First the cross-section, and polarization. That was the first paper by Palmieri, Cormack, Ramsey and Wilson, which you saw. Then, there was a measurement of "D" which was done by Hwang, Ophel, Thorndike and Wilson. They also reduced the beam energy by putting absorber in the beam, and measured "D" at 90-80 MaV. That was done by Thorndike and Ophel. "R" is the first experiment in which I worked, starting by helping Ed Thorndike to finish his thesis. And then, Ed helped us afterwards, and stayed on as a post-Doc for two years. And then "A" was measured a bit later by Thorndike, and Hee and "R'" by Hee and Wilson. So that was all the proton measurements

The cross-section and polarization for the proton deuteron intearaction were a series of experiments to test the idea that you measure the sum of proton proton and proton neutron interactions when you use the deuteron as a target. So, you had a series of experiments by Stairs, Wilson and Cooper. Kuckes, Kuckes and Wilson. Wilson and Postma, on cross-energy and polarization on the deuteron. And then, "R, N, A" were done by LeFrancois, Hoffman, Thorndike, and Wilson. And then, neutron-proton experiments were done. Polarization by Carroll, Patel, Strax, Miller, "D" by Carroll, Patel, Strax and Miller. "D" again, on a different angle by Collins and Miller.

Now a few words on the life and role of a Ph.D. student around 1960. It was a fascinating time. The Cyclotron Lab was our lab, and even sometime "our home". And this was not only because we spend a lot of time there. The usual shifts were 13 hours, and that would go on for 10 days in a row. But, it was our home. Because in those times, and this must not have changed, the winter can be pretty horrible, and the summer in Cambridge hot and humid. What may have changed was that undergraduates and graduate students were pretty poor in those days, and could not afford air-conditioning. And the only place which was free and air-conditioned was the lab. So, you would see sometimes the spouse coming, and people playing cards, reading books, and coming. Sometimes children in the coffee room, close to the Cyclotron machine. So, it was really a home. Now, the machine. We had really an excellent team to maintain and repair it. But, to get the Ph.D. we were expected to run it alone at night. Except that, for safety reasons, we had to have a baby sitter. So, we had undergraduate students which were sitting there with us, so that if we collapsed of a heart attack, because of a problem they could warn somebody. But we had to do also our ion source repair alone. And I think that I've taken more radiation in my Harvard years, than in all the time after. Because changing this ion source involved pulling out the probe, replacing the filament and the filament was pretty radioactive. We were also allowed to suggest modifications to the machine and try to make them. So, I'll give a bit of my experience on that. Because my thesis was doing scattering on this proton and neutron, we had to measure a coincidence between scattered particles on the two sides. We were limited by the random coincidence rate. Because the Cyclotron was ejecting the protons over a short time, something like 50 micro-seconds the dead time was a nuisance. In the slide are shown the teeth of the rotating condenser. (Picture in the main text) I had the idea that if you shape those teeth, maybe you can have a different frequency-time curve when they pass in front of each other. So, I took a bit of an aluminum piece, machined it in the machine shop, and imitated the section of the rotating condenser. And then, passed it slowly measuring the capacity. And, it turned out that by shaping it correctly, I could predict that the frequency-time curve would be better. And then, I went to Andy Koehler and asked whether we could do something about this on the machine. We had a spare rotating condenser, and we put it in the machine shop and changed the shape. It was done, if I'm correct, in a few months time and gained us a factor of 6 for the duty cycle of the cyclotron. I'm trying to imagine the face of the certain engineers now if an undergraduate student would come and say: "I have an idea on how to modify the machine for my thesis." We were really privileged at the time to have ideas like that. For the electronics, it was the same thing. There was very little commercial electronics. A pulse analyzer. And, most electronics was home-made with vacuum tubes. But the fast transistor has arrived in 1959. And, then we could design the coincidence circuit, gate and stretcher, fast-pre-skaters. And the help that we got, was of course discussing with the other graduate students who were there. And Dick Wilson generously paid some undergraduate student to help in the cabling. There was a competition with Bernie Gottschalk on who could design the fastest coincidence circuit. Bernie used the 2N-501 transistor. But, I got the 2N-1500, which had just arrived on the market. So it was a very good time.

Data analysis is not such a good story. There was no computer to do our data analysis. It was replaced by two undergraduate students, working on parallel, on mechanical machines. Now the younger people may never have seen a mechanical adding, multiplying and dividing machine. For a division you entered your two numbers, and it goes glock, glock, glock, glock, glock, glock for a few seconds, until you get your results. So, this was not my greatest assets to do the physics in my future. But all in all, it was a wonderful atmosphere. We were, what is, I would say is very different from what is happening today is that we were allowed to learn by our mistakes. I made some. But, certainly, this is something we all regret. The students now don't have the chance to build something and to say: maybe we'll succeed, maybe we'll fail. Nowadays you don't give students a chance to be wrong. We had Trevor Ophel from a small group of more senior physicists. Of course, I had Dick as my thesis supervisor. A lot of the physics that I knew I learnt from Ed. Thorndike who stayed as a post-doc. Working with Ed was a privileged experience of learning how to do experimental physics. I think if I go back, I don't think I've learned a lot about nuclear forces by doing this thesis. But, I certainly learned, and we all learned how to design an experiment, and how to perform an experiment. So they was great times. Thank you.

BERNARD GOTTSCHALK

I won't be talking about my recent work. I have worked on medical work for quite a few years now. But what I'll do now is just speak about the nuclear physics program during the physics part of the Cyclotron's activities. First of all, I'd like to thank Andy Koehler, for among many other things, keeping excellent records of Cyclotron activities. In the late '60's, there was a plan to write a final report for the Cyclotron. That never actually got finished, but some of the source materials are very useful. For instance, there's a list of about 280 publications connected with the Cyclotron. There are about 35 Ph.D. theses by Harvard students, and about 10 by outside students who used the Cyclotron. (That is, people from other universities.) And we have almost all of those theses. They are a lot of fun to read because, in a thesis, people tend to let their hair down. This, for instance, from George Gerstein's thesis. First of all, you see, it's actually a carbon copy. This is on flimsy paper, and it was one of three carbon copies. And it says "The adjustment of beam focusing conditions was carried out by placing a zinc sulphide screen at the desired target location, and observing the screen illumination with a 60-power telescope and two mirrors from a safe place." (laughter) So it's really amusing, to read these theses. Eventually though, after having fun reading the theses and some articles for a couple of days, I realized that I had really better start working on the talk. So, unlike what Jacques did, which was quite a systematic account of the nucleon-nucleon program, I'm really just going to give some highlights of stuff that appealed to me. Here's the first one. In the early days, the Cyclotron didn't have an external beam at all. And yet, many excellent experiments were carried out. They all had to be done inside the machine, and some of them used a great deal of ingenuity, such as this experiment by Norton Hintz and Norman Ramsey. This experiment wanted to measure the excitation function that is, the energy dependence of making certain isotopes, if you bombard a target. The standard technique, which (already at this time) was an old technique, and is still used today, is the stacked-foil technique. You just take a stack of foils which is deep enough to stop the beam. So at each foil, you know the energy of the proton. In other words, if you know the incident energy, then by using the stopping power relation, you know the energy of the protons hitting each foil. And then, after the exposure, you take the stack apart, and by various means – radioactive analysis, and so on, you measure the radioactivity. Today, it would be by accelerator mass spectrometry. You analyze how many isotopes you've produced, and that gives you the energy-dependence. Unfortunately, the technique does not work unless the energy spread of the incident beam is very small. Because a large energy spread would translate into a huge uncertainty of the energy at the end of the stack. In those days, we didn't have an external beam, and the energy spread of the internal beam was very large, because of the radial oscillations. Well, this technique overcame those problems. They put a small scatterer at one location. And that scattered a bit of the beam. And then 180 degrees away, they put their stack. It was outside the median plane, so that you wouldn't interfere with the protons that were en route to the scatterer. This way, they used the Cyclotron as its own analysis and focusing magnet. So it was really a very ingenious technique, which made the most of what they had. And, here, just for instance, are three excitation energies that they measured for daughters of Aluminum 27. So that was an example of an experiment done inside the Cyclotron. A few years later, around 1953, there was, in fact, an external beam. It was fairly crude, and the beam wasn't very intense. But a few experiments were done with it. This is from Walter Titus' thesis. He was a student of Karl Strauch's, and I second Dick's real regrets that Karl couldn't be here today. So, here was the external beam. There was a magnetic channel which got some of the protons out, and then some slits for energy analysis, and a bending and focusing magnet. And the result was an external beam at about 95 MeV. It had rather poor intensity which was the main motivation for the upgrade a few years later that Dick described. And here was Titus' set-up: just to give you a feeling for the style. He used a range telescope. A range telescope is an arrangement where you basically determine the energy of the protons by how far they go in a stack of scintillators. Remember in those days, electronics was fairly crude, and fairly expensive. So, just having enough coincidence circuits to implement this sort of arrangement was non-trivial. When you started taking data, I remember one spent half a day, or a day, just verifying that all the coincidence circuits were working, which is something you almost take for granted, nowadays. Titus and Strauch were interested in looking at elastic and inelastic scattering from nuclei. So, this is an experiment where the proton really interacts with the nucleus as a whole. It sort of sees the nucleus as a cloudy crystal ball. And it can either just bounce off of it – elastic-scattering – or, it can interact with it somewhat, and leave it in an excited state, which subsequently decays. This experiment, though, only looked at what happened to the proton, after it was scattered. And you can see, in this case, for instance, where the target is carbon, the proton either comes out with the energy corresponding to elastic scattering, here, or it can leave the nucleus in one of a number of well-defined excited states. In those days, a major topic in physics was determining the excited states, and measuring how they were excited by incident protons, among other things. And, for instance, this is also from Titus' thesis."Elastic Scattering as a Function of Atomic Weight". It increases rapidly, because you have here a coherent process which depends upon all of the nucleons. And therefore it increases more or less as the square of the number of nucleons. Well, then there was the shutdown, after which the Cyclotron came back up with three beams. And, I should mention, by the way, that the shut-down took just about a year, and really involved a lot of work. That is the longest time the Harvard Cyclotron has ever been shut down in its entire career. So then, among other things, you had a high intensity proton beam, at 160 MeV. And Karl's program was continued by George Gerstein, who was actually the first graduate student that I worked for. And I was one of the folks that checked his coincidence circuits to make sure they were all working. And his technique was not very different from Titus'. Here you have a range telescope. Now, there's a vacuum chamber around the target to try to reduce the background from air. Because of energy resolution considerations, the target itself has to be rather thin, and that makes the air contribution appreciable. Therefore we used vacuum chambers to try to get rid of the background from air. Otherwise, the experiment was pretty much the same as before. George was interested in measuring the angular distribution of elastic scattering, and got results like this one. This measures the angular distribution of elastic scattering from lead over several decades. The fit is basically a phenomenological fit with a nuclear potential that has both real and imaginary parts. By this time, computers were coming into use. The parameters of this fit were adjusted by using a Univac computer which was one of the first University-wide computers at Harvard. By the way, apropos of that, I want to tell everybody that the very first Harvard computer, or about half of it, is on display in the hall. Many of you have probably already seen it. It's the Mark I, and you should really take time to look at that machine. It was the world's first programmable computer. It did not use vacuum tubes. It's mechanical. But it had the feature that, rather than being built to solve a specific mathematical problem, it could solve a whole array of mathematical problems depending on a program tape that you fed in. So it was the world's first programmable computer. Have a look at it during one of the breaks. It was built towards the end of the War, and was used by the Navy for about three years. Mostly, I understand, to crank out Bessel functions. There's a very excellent set of descriptions and photographs under the computer. It's right behind where we were having coffee. So, actually, it's a little bit older than the Harvard Cyclotron. But, unlike the Cyclotron, it has not been in continuous use, (Chairman's comment: it was still working in 1955).

That's amazing. By today's standards, it is a 50-hertz, I think, 2 kilobyte machine. (laughter) So now I am going to get to some experiments that I did with Karl Strauch. Now, the experiments I've talked about so far were experiments where the incoming proton interacts with the nucleus as a whole. But, already in the early '50's, Chamberlain and Segre at Berkeley had established the existence of an entirely different kind of reaction, where the incoming proton would knock constituents out of the nucleus. The simplest thing is that it would knock another proton out, and at that time, the surprising thing is that the kinematics were such that it was almost as though this proton that was being knocked out was a free proton. And these things came to be called quasi-elastic reactions. And, in fact, from a nuclear physics of view, as you go up in energy, that's where most of the action is. In other words, the more interesting things to study are where you are knocking out nuclear constituents. Of course, history repeats itself 10 or 15 years later – at SLAC, they were doing "deep inelastic reactions on protons" and all that really is, is knocking constituents out of protons. So everything seems to repeat itself, over and over again, on a smaller scale. Well, anyway, in this case, we were smashing nuclei. And, I've never understood why these machines are called "atom-smashers". Anybody can smash an atom. What we were smashing was nuclei, and that's the harder thing to do. Anyway, to come to the apparatus. In my experiment, we had two telescopes to detect the protons. The principle of this experiment is that you would determine the energy of the proton by how much light it produced in a sodium iodide scintillator. We had found by experimentation that sodium iodide gave you the best energy resolution, and also the most linear response with respect to proton energy. Here's a close-up view of the scintillation counter. Of course, all of the counters had to be encased in iron shields because, even 50 feet from the Cyclotron magnet, there is still an appreciable fringe field, which would otherwise prevent the phototube from working. If you look at the lower picture here, it's interesting, because it shows some of the data logging equipment we had in those days. We're in the late '50s, early '60s and devices called "analog to digital converters" had just become available. There was a company in Cambridge that made them: they cost $5,000 dollars apiece. And they did use transistors, but those were very early transistors so you had lifetime problems. Occasionally, a transistor would actually fail which almost never happens nowadays. And then, you would have to go in and find out which one it was, and replace it. Well, anyway, we had two of these things to convert the two proton pulses into numbers. And the first thing we did was display the numbers on these neon read-out tubes, and record them on a movie film that was slowly moving. And then some poor soul, namely me, most of the time, had to read these numbers, transcribe them, and do the calculations. I got tired of doing that. And, by that time, IBM cards and card punches had come along. So at about that stage in the experiment, I built a huge matrix of relays that would drive an IBM card punch. And, every time an event came along, which was a few per minute, as I remember, the card punch would burp, and punch 6 columns. And, at the end of the experiment, I would have 10 large boxes of cards. I would lug them over to the central computer which, at that time, was at the Smithsonian, and it would process them. It was an IBM computer and it was programmable in Fortran. I always feel that IBM has not gotten enough credit for inventing Fortran, which was the major thing that really made computation accessible to scientists. So, that was the style in the early '60's. So this is an experiment where we knock a proton out of the nucleus, and then measure the energy of both protons coming out. And, the neat thing about this experiment is that by doing the kinematics, you can find out what the momentum was of the target proton before you hit it. So, in a way, it is a very direct measurement of the wave function of the target protons in the nucleus. These results I'm showing here are for carbon. For each event I've just done a scatter plot of the energy of one proton versus the energy of the other proton. And you see that it has a very clear structure. If you go in this direction, that is, along the diagonal – I don't have time to prove this to you – but if you follow through the kinematics, the coordinate in this direction is just the binding energy of the proton that you knocked out. So you see that there's a very clear group here of protons with one binding energy. And then, further down, a much fuzzier group of protons with another binding energy. And then, if you further do the kinematics, you find that going at right angles to the diagonal, that coordinate is equivalent to the momentum of the proton that you knocked out. So this group of protons seems to have a peak momentum which is not zero. It's greater than zero. Whereas this fuzzy group of more strongly bound protons has a momentum which is very broad, but peaks at zero. So this diagram, if you know how to read it, was the most direct demonstration by far, that carbon consists of four P-state protons which are weakly bound, and two S-state protons which are far more strongly bound. And these quasi-elastic experiments went on for quite a while. They were done by all our sister machines in other parts of the world: Orsay, Uppsala and Harwell. And really improved our knowledge of the nuclear model. Now, you don't just have to knock elementary particles out of nuclei. You can also knock clusters of particles out. And, a few years after the experiment that I've just shown, Sue Kannenberg, who is in the audience today, and who was at that time a graduate student at Northeastern University, did this experiment under my direction. In principle, it's very similar. We're knocking something out of the nucleus but in this case we're looking for the scattered proton in coincidence with an alpha cluster. So, this is called a $(p,p\ \alpha)$ experiment. I don't have time to tell you how we identified the alpha and all that sort of thing. But we see very clear evidence of an alpha cluster, with a very unique binding energy and a unique momentum distribution. The thing that makes the alpha experiment even more interesting is that unlike the proton-proton scattering cross-section which is almost uniform in the center-of-mass system, the p,p alpha scattering cross-section has this enormous momentum dependence. You see here, it varies by almost a factor of 30 over the angles that we explored. And yet, that very same momentum dependence that's measured for free protons and alphas seems to also apply when the alphas are bound inside a nucleus. So, actually, to an amazing degree, these quasi elastic reactions go on and behave almost like free interactions. And they give you a further insight into nuclear models. In other words, a nuclear model can now try to predict how often you will hit an alpha cluster and that sort of thing. Well, for the remainder of my talk, I'd like to concentrate on what was one of the last experiments at the Cyclotron. It was actually a nucleon-nucleon experiment. But, because I was the principal investigator, and we used a lot of the techniques that had been developed for the nuclear physics, I was chosen to give this part of the talk. The motivation of this bremsstrahlung experiment is the following. All of the experiments that Jacques described, this lengthy program, really, only measured cases where the protons scatter elastically off nucleons: where p,p or p,n reactions occur elastically. That is, energy is conserved. Now, if you want to use those results to interpret nuclear models, you have to cover cases where energy is not conserved in the two-nucleon system, because you're dealing with a many-body problem. I mean the simplest complication is when nucleons scatter off each other in the nucleus, they're bound. So energy is not conserved in that reaction. The jargon for that is off-energy shell, or off-mass shell scattering. And people were interested in measuring that. You can't do it really very well by studying nuclear physics, because there's so much going on in the nucleus, that it is hard to separate out these effects. And the thought was that it would be easier to separate them out, if you used a simple process. And the simplest process people could think of, was the emission of a photon or a gamma ray. And because it's the same as bremsstrahlung, by which we make X-rays for diagnostic medicine, it's called proton-proton bremsstrahlung. The desirability of doing this experiment had been realized for a few years. It is, however, a very hard experiment, because the cross-section is more than 1,000 times lower than elastic scattering. So the event rate is very small, and elastic scattering presents a huge background. So most of your ingenuity has to be devoted to trying to get rid of that background. Dick had written a book, which summarized the experimental knowledge about nucleon-nucleon scattering. and he was urging both me and Ed Thorndike, who had recently gone to Rochester, to try this proton-proton bremsstrahlung experiment. The technique we finally settled on was really an outgrowth of the nuclear physics.

If you do a scatter plot of the two proton energies, and what you had going on was a bremsstrahlung event, the two energies should lie on a ring-shaped structure. And the size of the ring depends on the angle at which the telescopes are placed. That arrangement subsequently became known as the Harvard geometry. In a preliminary run, we saw some signal. But the experiment wasn't nearly clean enough to say that we had seen proton-proton bremsstrahlung. We decided that we had to make a further improvement of the Cyclotron duty cycle, which we did. But that took another year. During that year, we got a progress report from Ed Thorndike, who is also in the audience today. It said "Dear Bill" – Bill Shlaer was the thesis student on this experiment – "Enclosed is a progress report on our work. A crude comparison with Cromer and Sobel's cross-section (two theorists who had tried to estimate the results of this experiment) suggests that they are high by a factor of five or more. Which turned out to be almost exactly correct. This letter, of course, got us very worried, because it showed that Ed Thorndike was hot on our heels. Or he would say a little bit ahead of us. But anyhow, we finally did the experiment in June '65. And, I'm really skipping ahead here. You can see even here in the raw data, the clear evidence of those rings that indicate that what is going on is proton-proton bremsstrahlung. The thing I forgot to emphasize is this is a way of detecting proton-proton bremsstrahlung without even looking at the gamma ray. Because large gamma ray detectors are difficult and expensive to build. So what we had invented, basically, was kind of a short-cut, whereby, just looking at the protons, we could identify the reaction. The experiment was done on-line to a PDP-1 computer, which had been set up and the software had been managed by Al Brenner. I should mention that because for its day it was very advanced. It was not only serving our experiment but, at the same time, two other experiments at CEA. The high-energy group was kind enough to let us use it as well. And it was what would nowadays be called a "real-time multi-tasking data collection operation" which, for those days, the mid '60s, was extremely advanced. It was certainly one of the first of its kind, with a 10K memory. We were allowed to use 2K of that memory, one tape drive, one oscilloscope output and one input port. Those were the resources we had in those days.

Just one final note which is tragi-comic. During our main data collection run, a tragedy occurred which was the CEA explosion. And, at the time there were just three of us to take data. So, we had hired this guy named Val Kirsis who was sort of a mercenary. He was actually a chemistry major. But he took data for a lot of Cyclotron experiments in those days. And was extremely conscientious. And, here, very early in the morning on the 5th of July, are his notes. It says "Beam off for a while. CEA explosion." At that point, he went out back. The explosion had blown in one of the doors, which set off an interlock, so that the Cyclotron turned off. So, he restored the door, and went on taking data. Then it says "Then stop run. Firemen on roof." Because the roof was a fairy high radiation area, we weren't allowed to run the machine, if there were people on the roof. So, as soon as the firemen came in, he conscientiously turned the machine off. Of course, a few minutes later, the firemen turned off power in the entire area so they could fight the fire. So, anyway, this is sort of an extreme example of what you find, when you go back through old data books. The CEA explosion has assumed almost mythic proportions in Cambridge. And, usually, one way or another, the Cyclotron is implicated, even though we had absolutely nothing to do with it. Just a couple of weeks ago, a friend of one of the Cyclotron operators was being shown an apartment by a real estate agent, and he asked "Why is Oxford Street closed?" And the answer was "Because the Cyclotron blew up." So I think what we have here is a very distant echo of that tragic accident at the CEA.

RICHARD WILSON

Just before I call on people I just wanted to mention a few people who have sent their regrets, David Bodansky, and Bob Birge. I will mention Owen Chamberlain. He was here on leave for six months, and was starting an experiment at the Cyclotron, which got interrupted. But he was on shift. And, actually having lunch at the Faculty Club with us, sneaking off and leaving the graduate student to run the machine, when he got a phone call from Stockholm, which made him the famous announcement. So, we went back to the Cyclotron and I arranged the first celebration of his Nobel Prize with champagne at the Harvard Cyclotron control room. At the control panel where he was sitting down, running the machine. But, he's unable to get here, because he also is not particularly well. Dave Measday is rushing off trying to sell his apartment at the present moment, otherwise, he could come. And we are not able to talk very much about his extremely fine work with a mono-chromatic neutron beam. I just want to mention three other people who have unfortunately passed away. Chester Hwang deserves a worthy note. He was a pilot in the Chinese Air Force. And went to Taiwan with Chang Kai Shek and then came to this country to get further training on jets. And then, his General absconded with $2 million dollars and went to South America, leaving him, leaving all the pilots without any salary. So, Chester Hwang came to graduate school. but the thing we should note, because it is in the news nowadays – what happens to Chinese (or Taiwanese) in this sort of position. Chester later got married, and had four children. And, sent pictures of his children back to his father and mother in Beijing. And he was very careful not to tell them that he ever got a degree in Physics. Or that he'd ever done anything in physics. Only told them of what he children were doing. And perhaps that's a note of caution that one should look at when one's reading the Cox report on the Los Alamos issues. Well, I think the first person that I'll call on is Lee Davenport, who is here and will tell us something.

LEE DAVENPORT

After all of this long-haired commentary, concerning nuclear physics, and protons – I've memorized all these words – and a number of other things. I thought it might be well to look back a little bit less formally, at what went on in the earliest days of this Cyclotron. I must say that I'm one of the very lucky people who's sitting in this room, for having had a chance to be associated with it at all. And my luck started way back, at a long time ago. It started at a meeting at MIT in October of 1940. A year before war was declared. That meeting was a meeting called "A Conference in Applied Nuclear Physics". There were some 600 key accelerator people who got together at the time. One of which was a man whom I'd never met before, named Kenneth Bainbridge. I was a graduate student then. I noticed that a number of the upper-level people if I might call them that, disappeared from these meetings and weren't there for all of the sessions. And, of course, the reason for that was that nuclear physicists were being summoned to what was soon to become the radiation laboratory at MIT. And, Ken Bainbridge was a key member of that initial group. Ken became kind of a hero for me. When I joined the radiation lab a month or so later, I reported to Ken for a brief period, but he moved onto other projects. I stayed there with the initial project that I was involved with called "The SCR-584 Radar Anti-Aircraft Fire Control". But the radiation laboratory brought to a number of us, what I would call a feeling or urgency. We were there to win a war. And we were there to get things done. And, Ken Bainbridge contributed a great deal to that feeling. And, I enjoyed the chance to work with him. Of course, he was siphoned off to go to Los Alamos, as you know. And, pretty soon, the Cyclotron here at Harvard was also siphoned over to Los Alamos. And, Ken didn't return until the end of the War. However, he and Curry Street got together and met me just as the radiation lab was closing and said "you know, we've got an exciting program here starting up at Harvard. We're going to build a Cyclotron to replace the one that's gone. And would you consider coming up and being a member of the Physics Department? And joining the Faculty?" That meant that I got to go to the Faculty Club and eat horse steak, which we served in those days. And I considered this to be quite an honor at the time. And, to become a part of the team and it truly was a team effort that led to the construction of this machine. By the time I got here, which was in September of 1946, Ken and Curry Street, and Roger Hickman, Bob Wilson had pretty much put all of the specifications of the machine together. And my job, somewhat ill-defined, it was a research officer on the staff, and the teaching faculty. I'm not surprised that the title of the exact job came through was "Coordinator," or, whatever it did come through as, because we weren't all sure of what our titles were. Everybody just settled in to work, and get this machine going with a sense of real urgency. The first meeting I ever attended of this group was with the architectural firm of Coolidge, Shepley, Bullfinch, and Abbott, who did all the architectural design for the buildings at Harvard University at the time. And I thought, boy, this is really a strange situation. The old Cyclotron sat in a wooden, World War I building called Gordon McKay. And now we're talking about building a concrete monument with another building alongside of it... with the world's leading architects. And that was my introduction to what this project was all about. We started by having to hire people. It was a stand-alone program. Had to have your own machine shop, your own engineering department. We couldn't depend on the rest of the facilities in the Physics Department. So, there you get the beginnings of what was called "The Nuclear Laboratory", not the Cyclotron laboratory, but the Nuclear Laboratory. The urgency was clear. Harvard was probably the only major U.S. institution offering graduate work in Physics. It did not have much of any equipment. The Cyclotron was gone. Much of the surrounding equipment was gone. And, the number of graduate students in Physics who wanted to study nuclear, or atomic work, had jumped from about 5 pre-War to about 100 Post-War. So there we sat with an increasingly large requirement for facilities and no machine. The urgency was clear, and we went to work on it in a great hurry. I never would have believed 53 years later, that I would be standing here, talking about a machine, which is just now ready to shut down. Now, at that time, Cyclotrons were growing pretty rapidly in size. The 184 inch was already under way. And unlike today's computers which become obsolete 3 weeks after you've bought it, it's up to another 500 megahertz. Like those, the Cyclotrons were moving pretty fast. But we knew that we had to get going on this. And so, most of the activity that went into the original building of this machine, was hurried, and rapid activity. We engineered as we went along, no question about it. Were we going to build the magnet out of 16 pieces or 4 or 17? And if we did, how were we going to get it machined? And who was going to cast the pieces? Obviously, the biggest problem of all was getting that magnet. And, between Bethlehem Steel, and Watertown Arsenal, which did the machining, and a company that we hired from California to move the pieces of steel from Watertown Arsenal down here, and rig them together and make a Cyclotron magnet out of it... were doing this job for the first time. In fact, they had to reinforce some of the culverts under the streets over which that equipment was brought, to get it here without breaking through the pavement between Watertown Arsenal and Oxford Street. Well, you've heard a little about these unusual things. I'll give you a couple of more tid-bits, and then I'll get out of the way, and defer to people who have used the machine, rather than helped to construct it. The basic specs, as I said, were pretty well done. We had, however, a lot of what we considered to be very clever ideas worked into it. One, was a movable shielding. Movable shielding was quite a new concept. And the idea of how to move it, was another new concept. We put in a crane over the top that would pull cables, and it would move the shielding when you needed to do that. Hopefully, that was also going to lift the magnet pieces in place. Of course, when we found out how big they were, it was clear that no crane was going to lift them in that building. So the rigging people had to come and do that. So, not everything worked smoothly. The stainless steel problem, which Norman so clearly referred to. We used more Glyptol than GE (General Electric) could manufacture, trying to get that thing leak-proof. And, finally, we heard about a company down in Long Island, and I remember their name was VECO… Vacuum Engineering – Leak Detectors. And we bought them. One of their first – we pried it out of them – to bring it up here to help us find leaks. Norman pointed out that probably the strangest of the leaks, the one that was in the vacuum pump itself. But, we did funny things. And another one that I'll conclude with, Cooling Water. Cooling water was going to be a problem. You couldn't dump it in the sewers. The City of Cambridge wouldn't handle that. We had to get rid of it. And if we had to get it in, we decided to build a well, dig a well, and, we did. We dug a nice well on the site. And, dug another return well to put it back in the ground.

EDWARD THORNDIKE

It's hard to know which things to pick out of the many Lessons Learned, and experiences had at the Cyclotron. And, I could, at the risk of wiping out the afternoon session, go on at length. So, I tried to phrase things in terms of where things had changed and where things have changed the same? What are the differences? What are the similarities today, versus back in the early '60s, when Jacques and I were doing a thesis. One of the things that has changed is the calculations that were being done for us on these manual computers. Not hand-calculators. "Marchant"s. They were being done by what Jacques really would have said were Radcliffe girls. But today, he had the wisdom not to say "Radcliffe girls." he had the wisdom to say "Harvard women." That is a assuredly one change. One of the things that probably has not changed, was that they were just as sharp then as they are today. In order to be sure that the calculations were done correctly, we needed to build in checks. And so, I had cleverly devised a way of making their calculations redundant without telling them. They were calculating the same thing twice in two different ways. And I noticed that one of them was doing these calculations pretty quickly. And I noticed that they were also coming out with no errors. And I said, "gee, again you're doing things pretty quick." And she pointed out, well, you know, I looked at the instructions you gave me, and saw that there was a redundancy. And so, I found a short-cut and so, I did not have to do the second set of calculation. I just copied the answer down first. There's a lesson in that – do not try to trick people who are smarter than you are. That person was Natalie Hubbard, who some 20 years later was a candidate for the Board of Overseers and was President of one of the major New York financial institutions. So, we had the advantage of some very, very smart undergraduates working with us.

One of the changes is that today, a grad student doesn't have any control over the machine. Then, we were responsible for the machines. But, that's sort of true. We were responsible for the machine, until we got into trouble. And, if we got into trouble, we ran to Andy. And Andy would take care of things. So, perhaps the advantage that we really had is we had a chance to watch a real pro in keeping the machine working. And, of course, a major strength which you saw instantly, was the large bank of wisdom by which he completely understood the machine. But the thing that came out, little by little, was a couple of other personality traits which were key. Extreme determination. And a great deal of patience. I remember, specifically, one February day in the winter, when the machine was just impossible. And so, I ran to Andy, "the machine's not working." Andy from about 10 in the morning into the evening, solves one problem after the other, after the other. Finally, well into the evening, Andy disappears into the night, going home for a well-deserved dinner. 10 minutes later, he comes back in the building. Picks up some tools. Goes out. Ten minutes later, he comes in carrying his car battery. After this entire day of fighting the machine, he has one more problem to solve, getting his car to start! I just could not understand how anybody under those circumstances, could have remained calm, and patient, and determined. I would have broken my foot, kicking the car. Or, I would have broken the windshield with a baseball bat. Determination and patience. I learned then, that it was the key to keeping a cyclotron working and a very useful traits for solving other problems of life.

I clearly have to mention something about Dick. I could not conceivably tell all of the stories about Dick. And some of them would be better if I didn't tell them. But there was one that showed me a lesson in scientific intuition. Where Chester, Trevor and I were starting our very first run. And Dick said keep the rates in the counters below a megacycle. Keep the rates in the counters below a megacycle. Otherwise, you'll have serious random coincidence problems. I did a little calculation. And I saw with the duty cycle that we had, and the coincidence resolving time that we had, that it was not going to be any serious random coincidence problem. So, we took the rates and the counters up above a megacycle. And the run was an absolute disaster. And the reason the run was an absolute disaster was that the high voltages sagged over the course of the spill. By the end of the spill, the data were worthless. And so, we'd wasted a month of our lives. And, I had to go to Dick, and to admit that we had not followed his advice. And, of course, I went to him and complained:

"You said that the random coincidences were going to be a problem but the randoms were not a problem" "Yes, but I knew you shouldn't have gone above a megahertz." I think that was the time when he told me this wonderful two-liner. "Learning from experience, means learning from mistakes." And I'm sure that any other of Dick's students could tell you the second line. "Learning from bitter experience, means learning from your own mistakes." I learned a lot at the Harvard Cyclotron. Much of it from experience. An awful lot of it from bitter experience. Certainly, you remember what you learn from bitter experience.

One is, in one sense, sad that the Harvard Cyclotron days are ending. But this is a photograph of Dick Wilson, with one of his students (myself). One of his grand-students Ron Poling (Professor in Minnesota) and two of his great grand–students. Steven Schrank, and Roy Wong. So, while there's great value in a child having interactions with his grandparent, there is also great value in a student having interactions with his grand-advisor, or his great-grand-advisor. And, both Ron, and Roy and Steve have commented to me about how much they're enjoyed having an opportunity to know Dick, and to talk to him, and to learn from him. So, at some level, the Harvard Cyclotron lives on through Dick's great-grand-students.

RUSSELL HOBBIE

It is the learning that we got at the Cyclotron is really important. As Jacques and Bernie talked about it this morning, it brought back a lot of memories. And I just wanted to share two of those with you. Putting in the neutron beam required installing 90 tons of steel shielding, which was delivered in the form of steel two by fours, about two feet long. And Bob Patel was reminding me that when these came, it turned out that they all had a burr on one end, where they had been cut with a power hacksaw at the mill. Our machinists refused to do anything about it. And so, for a week or two, Bob and Doug Miller and I sat at a grinding wheel outside the little shed at the back, grinding the burrs off. We really do learn great skills here! The other memory I wanted to share with you was about the first transistors and the coincidence circuits. Before transistors we had coincidence circuits that had an incredible beast in them, a Secondary Emission Vacuum tube, Phillips EFP-60. It came in a little can about this big, and about three inches long; that was the color of these chairs. And in the 1950s they were $100 bucks apiece. And, they were always the first thing to go. And so, in the electronic shop, there was a bin of fresh EFP-60's. You'd grab a couple. Put them in your pocket. Go out back. Just start a whole row of them, and the coincidence, you'd start swopping them out, and you'd find where the bad one was. And, we were sort of careless I those days, and we would throw these things on the floor. One night I was running, and the coincidence circuit stopped. And I went through the rigamarole. It never worked. So, then I checked all the co-ax, and all of this stuff. And, finally, in the morning, somebody came in who was thinking a little bit more clearly than I was at that point and said, maybe I had better try checking out the EFP-60's again. And, so I did, and it worked. Well, it turns out that the janitor we had in those days was very conscientious. He had heard us talking about how expensive these were. So, he was picking them up from the floor, and putting them back in the bin in the electronic shop!

SUSAN KANNENBERG

Well, Bernie was kind enough to refer to my work, which I appreciate. I didn't have any Radcliffe students to help me reduce data. We only had men running around the place including the baby sitters who definitely were not women. But I had two kinds of casual, or just little anecdotal or nostalgic recollections that I thought were… have stayed with me for all these years. One was one of the many times that we were running, doing, taking data. And our usual arrangement was that I would work 12 hours, 9 to 9. Bernie worked 9 to midnight. And the mercenary worked from midnight until 8 in the morning. And so, of course, even though we ran for three weeks, typically. And quite a long time, our event rate for PP-alpha was way, way low. When you had a problem, and you needed some work done to fix something, you really did need it to get done. And, in 1967, the Red Sox were making their rare participation in the last game of the World Series. And, I had a terrible time trying to convince people in the so-called Cyclotron Lab; I think that's what it was called. Not the Vacuum Lab. To tear them away from this important activity in order to repair some of the equipment that needed repair. So, I can remember that the Red Sox were in the very last game of the World Series. Because it happened in '67 on my watch. Another recollection that I have which is a little indirect from the Cyclotron, is that we were visiting some friends in Rome one time, and Dick happened to be there. What's a recollection without Dick? And we went to Frascati (Italy) where my friend happened to be on the staff. And, I don't know if she was not available, and, we met Dick there. And he was going to bring us in, and show us around and so forth. And we got to the gate, where there was a mechanical or electronic arrangement. You press a button, and voice comes over and asks you to identify yourself before they open the gate to let you in. And he identified himself, as you may imagine, in a most creative way. He identified himself as being someone on the less favorable side of the law, associated with The Family. But they let him in. Which I guess they knew it was him.

Anyway, I guess I had one other story if Bernie doesn't get too angry with me. I think this one is just really pretty funny. I worked very hard and diligently, I think. And I was somewhat intimidated a lot of the time, because I wasn't an electronics whiz by any shakes although I made up for it in other areas that took a while to convince Bernie. But I think that he eventually came around. But, anyway, one day I showed up and the place was just totally filled with black smoke. Does anybody remember that? Absolutely filled with black smoke. From the basement, all the way to the top. And as we were sort of threading our way through this invisible mess we saw Bernie laughing and saying, how hilarious this all was. He had just kept cranking up the high-voltage on one of the amplifiers, for one of the photo-detectors. He just cranked it up until finally it just burst into flames. And there was this huge fire in the back there. I was appalled. The place was just completely filled with black smoke. And you couldn't see anything. And, I asked Bernie, "why do you think this is so funny? I have done far less to make a problem, and it's really much worse." And he replied, "because I can fix it."

ALONSO

I'm Alonso. I'm one of the outside people that came and used the Cyclotron. In the mid-'60s, I was a graduate student of Lee Grodzin's at MIT. And we were involved in hyperfine interactions worked. And one of the things that I needed to do was to make radioactive sources. I needed a source of 50 MeV protons. And I had been getting sources from the Cyclotron at Berkeley. But, in fact, when Lee mentioned well, you know, there's a Cyclotron right up the way here. Why don't you try to see whether or not this, you could use it for your sources. And so, I came up and met Andy. And Dick Wharton, also who was the operator, the technician at the time. And they set me up to do some irradiations. And I have two anecdotes that I'd like to tell about this. That well, just to sort of finish the introduction, if you will, is that in a sense, this really represents yet another use of the Cyclotron, in different areas, even though I was involved in hyperfine interactions. The study that I was doing was really a solid state study. Namely, the measurement of internal magnetic fields, and super-conductors. And, I think it was one of the first measurements that was actually made of bulk presence of flexoid distributions in Type II in niobium. And, but then my association also with the Cyclotron through the medical programs. Because I've been at Berkeley for many years, and was very heavily involved there as the Director of Operations for the Bevilaq and Medical Programs. And so, I had a lot of historical connections with a Cyclotron. But, my first connections with them were really with the apparatus what was used. And the first anecdote is sort of the comment that Professor Ramsey made about training, and letting students using them (the cyclotrons). That, in fact, yes, that was the case with me too. Dick Wharton told me, here you are going to run over night with this thing. And, he set my target up on the internal probe. And there was a coffee can sitting in the side of the vault, which was an ion chamber. And there was a co–ax cable running out to an electrometer that was sitting on the console. And, my training consisted of five minutes. Namely, turn this knob, until that meter goes to there. And, that was basically it. So, that's all I knew about the machine. My first exposure to Cyclotrons… or to accelerators. But, as a tribute to the reliability of the machine was that yes, in fact, it ran all night without any problems. And I certainly would have been in hot water if there had been any problems. If I'd turned that knob, and the meter didn't go up at that particular point. So, I made several sources that way – and actually was able to finish my thesis with the activity produced on this machine.

But the other anecdote also relates to things that Bernie mentioned, associated with the CEA explosion. That during one of my visits up here, I was given a tour by a graduate student. I don't remember who it was, but who was showing me his apparatus on the scattering table. And it was sitting up in the back comer. And, he was commenting about a particularly unusual experimental problem that he had – because one of the effects of the CEA explosion was that screens on the very high windows, in the high-bay area, which were always open because that was hydrogen venting had been knocked off by the explosion. This allowed the pigeons to come in. And so, the problem that he had was to make corrections on his target thickness for pigeon droppings. Anyway, that's a tribute to the Cyclotron for its multi–faceted use. And, I certainly have enjoyed my interactions with Andy and Bernie, and Miles, and everybody. Thank you.

LEV GOLDIN

I wanted to congratulate the people from Harvard Cyclotron with their work on proton therapy. And I wanted to tell you several words about how it was done in Russia. So, we were demobilized after the first War, and had to begin a new field of Russian physics. No one had any experience in it. But we understood that it must be done. And, after several years, our heavy water moderated reactor. Then we thought that we could do really everything. We were quite young men, and didn't understand many of the difficulties which were before us. We thought it was possible to make a strong focusing accelerator. If so, why not make one? And so three of us began this work, and after several years, it began to work. It was not so simple. Because there were only two people who understand everything. There were some engineers and mechanics in radio-technics, in electricity who could work with us. But no one understands the problem as a whole. So we were told, yet we did it. And after it was done we understood that we can do really everything if you want to. And that was a time when Professor Pomeranchuk one of the leading theorists of Russian nuclear physics was ill. And, he invited me into the oncological centrum and he wanted to discuss with me, why it was that he was irradiated with electrons. And we thought about the problem, and found that it was not the best way. The protons could be much better. We didn't know anything about work done here in Harvard, or in Uppsala. We didn't know everything. You see, the connections between Russia, and all other world were closed. And, the only source of our knowledge was scientific journals, which we could read, if we could get them. And so, I began to think about it. And, after several years, the proton irradiation of patients began. It was successful as well. And only then did we know that we were not the first. That the problem was really an old one. And it was decided in Harvard. And it was half-decided in Uppsala. And that we began the work, and now our statistics, I think, are the second best statistics in the world. It is not so simple now, because we have no money. And, if our great accelerator when we began our work is really without money, the only possibility for it to work and to use it is because of proton therapy. Because it is recognized to be useful. I wanted to congratulate once more the people from the Harvard Cyclotron who were first to understand the problem. And who were so successful in doing it.

JAMES NIEDERER

People don't come back just with memories of a machine. One of the really beautiful things of this operation was the wonderful people at Harvard. Our mentors at every level. We remember what we all learned in the machine shops to start off with. It wasn't just a place to fix cars and things, which it had its share. These guys were good. They took us by the hand, and they got us through some very complicated stuff. In my case, I came in from some modest place. We were playing catch-up then. I guess I have been all my life. But still, but we mingled. And now, we've got a big national laboratory. We are treated like children for some GS-zero teaching us about some trivia! The faculty were friends. At least the people we worked with. They'd visit us in our dorms. We'd visit them at their homes. We shared the marriages of their children. Some of whom became friends of ours in other places. And, this is somewhat unique. I mean, partly this is what Harvard did for the Physics Department as well. And, the Cyclotron itself is a very special place. Harvard, in its great wisdom, gave the Cyclotron, effectively, to two people without tenure. Dick, and Karl Strauch. This gave a little incentive, I guess, to do a little more. And it took awhile before Harvard did relent and let you medical guys on in this. But in a sense, the students had to compete. So you got sort of reasonably good habits. And so, yes there are vignettes we can talk about. How we learned to use the crane. It's just that somebody happened to have left the beam on at that night. So it may have had a few more dents on it, which as near as we can tell from one or two of the parasites, when you try to run it the next day. Schlumberger gave a Fellowship to Harvard which, I guess, still goes. There are other wonderful connections here. I think we came back because of the people involved. We see our friends here… and this staff has contributed to the wonderful people that brought us here.

ROBERT SCHNEIDER

I hadn't intended to say anything. But I was inspired by the others. My name is Bob Schneider. I'm presently at Woods Hole Oceanographic Institution. Yes, we have an accelerator at Woods Hole, the only one on the Cape, actually. But, I was a student of Allan Cormack at Tufts, as Sue was from another institution. I want to represent our feeling from other schools of the generous sharing of resources that Harvard made in letting graduate students from other universities pursue research here, over the years. But, I can probably also add a couple of things to some of the anecdotes. I know that in Ted Howard's thesis, there was actually a correction made for the pigeon droppings, I believe. And, when Sue got up and spoke, I remember that at the first Harvard Cyclotron closing party 20 years ago, there was a telegram read from Sue, who was somewhere in the Caribbean unable to attend. But, I worked with Allan Cormack, and David Measday who had a neutron beam set up at the Cyclotron of about 150 MeV. And we measured a number of neutron cross-sections on heavy elements at that time.

Another sideline. I think I know where the term "atom-smasher" came from. There was a famous cartoon in one of the Bay Area newspapers. You know, before the War, when the Berkeley 184-inch magnet had been constructed, but was not yet complete. There were a lot of pictures and cartoons of it. It looked very much like a large press. And some of the cartoons showed it as being something that would squash things. And I think that's where the term "atom-smasher" term came from. Andy Koehler and I, have been informally collecting information on the other synchroc Cyclotrons of that generation. As you know, there were a number built in the U.S. And, all about the same time. And, some of the interesting developments that then went on to become part of strong focusing machines later on. And, I think, makes an interesting story. So, if I contact any of you for reminiscences or anecdotes, that's what it's all about. Thank you.

RICHARD WILSON

I want to get some things on the record of how much the Physics Department had done for the medical work on this machine. In 1967 Bill Sweet, who unfortunately can't be here, and Ray Kjellberg also for whom it would be more difficult to be here, had treated. roughly about 108 patients who had various pituitary problems on the Cyclotron. We all thought what they had been doing was impressive. Andy Koehler, and Bill Preston, who was at that time Director of the Cyclotron and Chairman of the Physics Department at that time did a great deal of help. They got a building paid for by NASA for doing various medical studies. And the particular excuse for getting that building was that 150 MeV was about the peak of the spectrum of cosmic ray particles, and NASA did not yet understand what 150 MeV protons did to people.

So, we, the Physics Department and I in particular, as Chairman of the High Energy Physics Committee liked the program. But our contract for running the Cyclotron was coming to an end. And we were about to write the final report, which we never did. The high energy program was bursting at the seams and needed space. We had a highly enthusiastic high energy physics group. We not only had an experimental program at the Cambridge Electron Accelerator next door, but also experiments at Brookhaven and elsewhere. We had all the 4th floor of the Lyman Laboratory where Margaret Law had a scanning facility for bubble chamber physics photographs. We had already taken over a building called Palfrey House and Geology Building. We still needed more space. And so, the obvious place to look for was the Cyclotron building. Since the cyclotron was going to be closed down, that made sense. But we didn't really want to close off something, if somebody really wanted it and could pay for it. So Bill Preston and I said to ourselves: "Who wants to take over the Cyclotron, and run it and make sure the bills get paid? Was it MGH? Or Kjellberg or Bill Sweet? The Medical School? Was it the Physics Department? Who?" Because ONR/AEC and Harvard were not going to be paying the bills any longer. Harvard University's well-known policy is that every tub had its own bottom. And, we were a tub. And we had to float.

So I went around and talked to the Director of MGH who was not helpful. In spite of Kjellbergs excellent work, the hospital staff did not think particularly well of him for personal reasons, I suspect. I went to talk to Dean Ebert of the Medical School. After two hours in his office, didn't even say no orally, and said he would write me a letter. I never got one. But I got my letter to him in my file. I wanted to be sure that he hadn't misunderstood the attitude of the Physics Department. We liked the work, and we hoped that we could be re-opened. "We were happy with any administrative procedure which is satisfactory to Harvard. Through MGH, the Medical School, or whatever. Financially, Mr. Kellberg believes he can find a fair portion from his sources. If Professor Hellman is also excited, presumably it would become a financially viable operation." Then Bill Preston cheated, which was probably the right thing to do. And, he pretended to Harvard that MGH would give more approval than they actually had. And he got an arrangement where they would indeed pay the bills, bit by bit, as Harvard presented them. And, Bill by that time, was the Director of the Physics Laboratory. So, he could juggle the accounts over here. In my note to Bill at the time I commented that it was a tremendous pleasure to all of us that Mass. General Hospital has agreed to continue to run the Cyclotron, and take responsibility thereof. But they hadn't! It seemed to me at the time a shame that an activity for which the patients were willing to pay should not be able to continue.

We found out about a year or two later that MGH wasn't as completely committed as we thought they were. And so, we had a second round of thinking about closing it down. And, various things happened at the same time. One of which, in my life, was unfortunate. The Cambridge Electron Accelerator was closed. And we could no longer say that the high energy physics program needed the space. The day after we got the bad news that the CEA budget was being cut form $4 million a year to $1.5 million a year, I told Ray Kjellberg it looks as if you'll be able to continue awhile longer. Then Herman Suit came and we never looked back.

About that time, we saw the paper about which we heard from Dr. Goldin this morning. Tobias and Laurence had started proton therapy in Berkeley. And, it was being done in Sweden. An interesting question today is why has Berkeley stopped? And why Sweden hadn't done as much as we have? HCL has performed an enormous number of treatments, and people have copied our treatments. In spite of the fact that Berkeley had a lot more funding than we have. Their machine was paid for by the High Energy Physics Program. In addition, they had an AEC contract for some of the medical work, and we had none. And I think that in ensuing talks that you will have this afternoon, you'll see why our Cyclotron has been so successful. It has been a matter of extraordinary pleasure to me, and I'm sure to Andy. There is no doubt, whatsoever, in my mind that Bill Preston and Andy were the primary people responsible for making the program work. One secondary factor is that we have had an extraordinary close cooperation with all parts of Massachusetts General Hospital. I think that we have succeeded in that far better than any of the other facilities who have been using radiotherapy. I believe these two factors are the secret of our success. And I think that this will be proved, as we go on this afternoon.

ANDREAS KOEHLER

Dick Wilson, in a very few minutes, covered everything I'm going to cover again in many more minutes. So, we apologize, both of us, I'm sure, for the redundancy. I wanted to talk a little bit about people. One of the important people in this field is Bob Wilson, who has been mentioned a number of times. Another important face in all of this is Bill Preston who was Director of the Cyclotron Laboratory and my boss. He hired me very soon after being hired himself in 1953, and we enjoyed working together for very many years. The loss of that companionship when he died a few years ago is still very much with me. He was Director at HCL form 1953 to 1975, which was the crucial period for the start of medical work there. I think you can tell even from this rather fuzzy picture that he was a very nice person to get along with. That was very important for getting this business to go, because, as Dick Wilson pointed out, cooperation between different ends of the University, and different parts of Mass. General Hospital, was crucial.

In 1958, as nearly as I can reconstruct it now, Bill Preston was approached by Bill Sweet and Ray Kjellberg from the Neurosurgery Department. They had been reading reports from Berkeley and Uppsala about initial use of proton beams there. Bill Sweet was quite excited about it and was pushing that something of the same sort should be done by Mass. General. He made a few inquiries and discovered, to his surprise, that there was a cyclotron just across the river that might fill the bill. So he went and talked to my boss, and got him excited about it too. Bill Preston called me into his office one day: would I consider working with him on developing this? Sure, it sounds like fun.

Now, what did the Cyclotron Laboratory have to offer at this point? It wasn't much. There was a cyclotron, there was a beam which was being used for physics experiments, and that was about it. There was a big machine shop, and there was a control room to run the cyclotron. And there was an awful lot of stuff out here connected with the physics research program. The problem was to figure out how to implement this idea of medical application of the proton beam. Well, there were some things that had to be built, and there were some things that had to be studied and understood. We had to figure out how to measure this kind of thing, and we also had to find room for it. The initial work was done in the Physics area, right there, with this external beam, and it was pretty primitive.

The beam came out through the shielding wall, from the cyclotron. I built a Uni-Strut frame to support a collimator. The colllimator was pretty primitive too. There was a brass plate there, and another there, with holes in them for the beam to get through. And there was a stereo tactic frame which could support, in principle, a patient's head so that we could irradiate the pituitary. And there was an x-ray machine salvaged from somewhere, so that x-ray pictures could be used for alignment purposes.

Now the plan that we worked out with Drs. Sweet and Kjellberg was to combine a rotational therapy, which was being used at Uppsala and Berkeley, with the use of the Bragg Peak – the stopping beam – which Bob Wilson had explained in his paper, years earlier, but which had never been used. The folks at Berkeley had decided it was probably too tricky. They couldn't figure out how to control the depth of penetration. So we concentrated on that, and came up with the scheme of using a telescoping water absorber.

This is a little hard to visualize. This water column can be lengthened or shortened, and a detector at its output end can measure the amount of radiation being deposited. If you do that, you trace out this Bragg Peak curve, which is characteristic of a proton beam. OK, we measured it. Now what do we do if we want to stick a patient in here? Well, the idea is that if you can adjust the patient's position, so that the center of the target coincides with this same point, then the Bragg Peak will always fall at the center of the target. And if you rotate the patient to a different position, the telescoping water absorber will compensate for any change in the amount of tissue the protons have to penetrate to get to the center of the target. It works to first order because the tissues are nearly water equivalent.

Of course, it is not all soft tissue: you have to go through the skull bone. We worried about that quite a bit, and made measurements on quite a number of specimens of skull bone, to determine the amount of slowing down in bone. We developed a system for compensating for that. And also, it is not necessarily true that you want to have the Bragg-Peak at the center of the target. If you have a target that's elongated, or a funny shape, you may want to use it a little bit differently. All of that was studied and pretty well understood.

[Andy showed a picture of Dr Kjellberg similar to the one shown on page 41.] I wanted you to have a chance to see him at work, because his concentration and individual attention to everything that was going on was a very important factor. It certainly was for me. And it was also of great practical importance, because what we did first was a lot of animal exposures to make sure that we really knew what we were doing. First we used a monkey brain, the monkey having been irradiated with the proton beam, using the water telescope arrangement to control the depth of penetration. We did this in a couple of hundred monkeys and compared the location of the Bragg Peak lesion with what we had intended. That gave us a pretty good measure of our precision.

We also learned some radiobiology in the process. You can see that the lesion – that smudgy, dark area in the monkey brain there – conforms reasonably well to the high dose region in the graph below. Ee studied that in great detail and discovered that one can actually see a change in relative biological effectiveness along the track of the beam. It's not a big effect, but it is certainly interesting.

All these studies allowed us to work up to designing a frame for supporting a patient. You notice this frame is similar to the one that I showed in the very first slide, but it's been beefed up in various ways. I think that you can just barely make out the bony fixation: the drill points that are set into the patient's skull bone to make sure that the head can't move unless we want it to. But, at the same time, there is a provision for rotation around a vertical axis as well as a horizontal axis, which in this case goes through the nose and the back of the head. That way one can approach a target from several different angles.

That equipment was used to start treating patients in 1961. By July 1974 we had treated over 600 patients. Most treatments were aimed at the pituitary gland, but a few targets were elsewhere within the skull. We had only treated 10 arterio-venous malformations by 1974, but that eventually became one of the more interesting targets that were explored. They have been treated a lot in other centers, following our lead at Harvard.

Here is a graph of a cumulative total of patients treated at Harvard. It starts back there in 1961. This is a log scale, which I love dearly because, you know, the first patient is very important. The second patient is almost as important. The 10th patient is almost as important as the first patient. And the 100th patient is about as important as the first 10. So that's why a log scale works pretty well for me. We had treated a lot of patients in the first 10 years. I'm also showing here treatments that were not done by Kjellberg in the Neurosurgery Department: the beginnings of Herman Suit's program in 1975, and Dr. Gragoudas's eye treatment programs in 1976.

I want to go back to reciting the history. It's hard to imagine making this talk really exciting, because you necessarily have to switch back and forth between matters of history and trying to show pictures of what goes on now. I wanted to come back to this because of the kind of patchwork arrangement of support that we had. Dick Wilson alluded to that already. We managed to get sort of under-the-counter support from the Office of Naval Research. The agreement was that it was OK for us to use the cyclotron, provided that did not interfere with the Physics research that ONR was interested in. It was 'non interference funding' that we got. Well, it was enormously important.

So here, in 1961, the first patient treated by the neurosurgeons. In 1963, we got the Space Agency to put up a chunk of money to build an annex so that there could be a separate treatment area. That was of interest to them because they wanted a place to set up equipment that they wanted to test. Then, in 1971, we got a grant from the National Cancer Institute – Mass. Eye and Ear and the Cyclotron Laboratory – to develop eye treatments. At that time we were talking about treating retinoblastoma. Well, we didn't get around to treating retinoblastomas until many, many years later. But it made a pretty good proposal, and we got money that way.

We also got money from the National Science Foundation – the RANN (Research Applied to the Nation's Needs) program. We argued that developing better radiation therapy was a national need. So by 1974, we had used all of this patchwork funding to put together something that was capable of treating a patient with a larger sized lesion. (I may want to go back to that.) There's a lot of the nitty-gritty of putting together funding implicit in that slide. But, certainly, the interest in learning how to treat larger targets than we were doing with the neurosurgeon was very strong.

For the most part, we were using single Bragg Peaks in the neurosurgical program. But, of course, you can offset. You can have one Bragg Peak penetrating the full depth, and then you can pull it back a little bit by interposing some absorber. And, if you do that with just two Bragg Peaks, you can get a flat top on the dose versus depth curve. If you do it with several Bragg Peaks, you can increase the extent of that flat top. In practice you do that by twirling a range modulator wheel across the beam, so that the depth of penetration is modulated. This was not new: it was in Bob Wilson's 1946 paper. Anyway, we worked on that and got it to work pretty well. We also had to worry about spreading the beam laterally. Originally, and we still do it quite often, we used just a single scattering foil, which gives you a Gaussian intensity distribution. If you're back far enough away from it, it can be a big broad Gaussian. But that is inefficient.

One idea for an improvement was to use a double-scattering system. The first scatterer gives you a Gaussian. And then, a little way down from that, there is an occluding plate or annulus that knocks out some of the intensity that's being transmitted. That puts a dip in the lateral intensity distribution. Then a second scatterer fills in the dip. For the sake of completeness, this also shows a large range modulator wheel, an ion chamber to monitor the transmitted beam, and a water tank to measure what's going on downstream. [This is shown in the top figure on page 43.] The first scatterer gives you a Gaussian. You stick in the annulus: the occluding plate. That knocks holes in the wings of the Gaussian. And then, you scatter it again and allow it to recombine, and you get a nice flat top, somewhere further downstream. That was all implemented with this patchwork of funding and support. We had to do this in the same beam line that was being used for the neurosurgical work, which meant that everything had to be dismantled and put back up. The annoyance of doing that emphasized the need for getting more space.

So then we took what used to be the high voltage supply room for the cyclotron main oscillator and turned into a treatment room. In order to get the beam into that room, and still be able to have it large, we had to cut a hole through the wall. The largest drill we had was only, I think, 8 inches in diameter. So Miles Wagner and I had to cut chunks out repeatedly, like Christmas cookies. I think it's worth noting again, the way that we have tended to work in teams that have a lot of affinity.

Meanwhile, a lot of people were paying attention to what we were doing here. As a result, other facilities got into the medical treatment mode. This is a cumulative count of facilities that were offer proton, pion, or ion beam treatments. [The list, as of January 2003, is on page 55.] If you look forward from

1961 let's say, when we started treating at Harvard and go forward another 30 years there are now 20 facilities around the world doing similar work. I think that is some measure of success.

There are other things that weren't so successful. Here's my hand bunched together so that all of it could be irradiated in a single field of the proton beam. And, you can see in the X-ray image up above, which shows the defining aperture for the proton beam, that pretty much all of the bones in my hand were involved in this irradiation. The idea was to try to activate the calcium in those bones, and from the counts accumulated, to get a measure of how much calcium there was in the bone. So, there's me sitting there for a very long time, being counted. Well, it works, but it was not a tremendous success. It's still talked about once in a while as a possibility. I think that Jim Adelstein is still an enthusiast for the technique.

I come back once more to the nice picture of Bill Preston. And to imagine for myself, and perhaps for you that he is pretty happy with what we accomplished. Thank you.

So then we took what used to be the high voltage supply room for the Cyclotron Oscillator and turned into a treatment room. And, in order to get the beam into that room, and still be able to have it large, we had to cut a hole through the wall. And the largest drill we had was only, I think, 8 inches in diameter. So Miles Wagner and I had to cut chunks out repeatedly, like Christmas cookies. I think it's worth noting again, the way that we have tended to work in teams that have a lot of affinity.

And, here's my hand bunched together so that all of it could be irradiated in a single field of the proton beam. And, you can see in the X-ray image up above, which shows the defining aperture for the proton beam, that pretty much all of the bones in my hand were involved in this irradiation. The idea was to try to radiate the calcium in those bones, and from the counts accumulated, to get a measure of how much calcium was in the bone. So, there's me sitting there for a very long time, being counted. Well, it works. But it was not a tremendous success. It's still talked about once in a while as a possibility. I think that Jim Adelstein is still an enthusiast for the technique. I come back once more to the nice picture of Bill Preston. And to imagine for myself, and perhaps for you that he is pretty happy with what we accomplished. Thank you.

KRIS JOHNSON

I get into the business of charting the growth of the medical program, more by accident than design. Back in mid 1970s we had already treated about 100 patients, in the Uveal Melanoma Program, and the Radiation Oncology Fractionated Program. And, it soon became obvious that we needed some kind of a database to answer basic questions, like how many patients we have treated. The various categories? How many doses? What's the dose they had received? Fractions, age and so forth. It was coincidental, at that time, that located just in back of the Cyclotron lab, is a building known as the Dunbar Lab. And, in that building was a mainframe computer, Digital Sigma-VII VAX.

I was quite familiar with the VAX. I had used it quite often when I was at the bubble chamber group over at Lyman Laboratory working, when Margaret Law was the Director. And it was coincidental also that even though I was working at the Cyclotron, I was still keeping the accounting records of the use of Sigma-VII. And, every month, I would send over a list of the users for billing purposes to Margaret Law. I don't remember whether I gave myself an account on Sigma-VII, or whether I just went in early, before everybody else got there, and worked on the machine. In any case, I wrote a Fortran program, and I entered the various patient parameters on IBM punch cards. And fed those into the card reader. And lo and behold, came an output. And, believe it or not, I'm a sentimental person. And I have kept the output. And, even more remarkable, I could find it. (Kris showed two of the original lists.) One is chronological, and one is alphabetical Dated August 15th, 1978. And I had even labeled it, which is even more remarkable, and it says "Original patient records, using Sigma-VII computer at Dunbar" and, I put that on a good old IBM punch card. That was the start of the record-keeping. I don't remember how long I kept on doing that. But one day I received a call from Michael Goitein over at Mass. General. And, he said that they had purchased a software program, again, Digital called Data-trieve. And I was welcome to use that to enter the patient information. I'm still using Data-trieve, just like the Cyclotron; I find Data-trieve very reliable. And, all of the numbers, the number-crunching you see here, comes from Data-trieve. Of course, things have changed over the years, we don't have the large output any more. We now have the good old 8½ x 11 page. And the small type that you can hardly read.

Here are all our patients. About, 4,600 of them since since 1974. A summary of the HCL Medical Program, showing the relative percentage contributions of the three major medical programs. We did treat our first patient in 1961, under the direction of Raymond Kjellberg in the Neurosurgical Department. And, he provided 100 percent of our patients up until 1974. In fact, he had treated approximately 700 before the Radiation Oncology Fractioned Program began in January of that year under the direction of Dr. Suit and Dr. Goitein. Approximately one and a half years after that in July of 1975, Dr. Gragoudas treated his first uveal melanoma patient. And, as of December 31, 1998, uveal melanomas have contributed approximately 25 percent of the patient population. Radiation Oncology, 50 percent. And the Neurosurgical Program about 25 percent. The Neurosurgical Program is now under the direction of Dr. Chapman, and Dr. Loeffler.

For the purpose of clarity on the graph, I have not included Dr. Kjellberg's contribution from 1961 until 1972. You can see that he treated, on the average, about 100 patients per year. There was a peak in 1985 of 230 patients. Since that year a steady decline in the number of patients, until1993. Dr. Kjellberg died in December of that year. Dr Chapman is continuing with the Neurosurgical Program. He began in 1991. His Neurosurgical Program contributed 139 last year. The next Program to start was the Radiation Oncology

Fractionated in 1974. For many years, we treated less than 100 patients per year. But starting in the early 1990's this increased to over 100 and in the last few years increased substantially, due primarily to the contribution of our macular degeneration patients. The uveal melanoma have been averaging over 100 patients per year. There were two years, 1985 and 1986 when we treated 194 for a total of 2,602. Our entire patient history, as of December 31st, 1998 we had treated 7,942 patients. Dr. Kjellberg had contributed 2,929. And, the "Star," which stands for Stereotactic Alignment Radio-Surgery, Program, is the continuance of the Neurosurgical Program and contributed 449 for a total of 3,378. The various categories underneath the radiation oncology fractionated, were chosen some years ago, quite arbitrarily. You will notice that the largest category here is the chordoma – chondral sarcomas of 636, for a total in this program of 1,962. And we treated our 8,000th patient in February of this year.

Some years ago, we became interested in the geographical distribution of our patients. This was before the NPTC was considered, to give an idea of where our patients were coming from, so that one could plan for potential proton facilities in the future. Of the 1,962 radiation oncology fractionated patients, 83 percent of those have come from the United States. All States except North Dakota, and Wyoming have been represented. 17 percent came from foreign countries. Of 2,602 uveal melanoma patients,. 94 percent have come from the United States, except the States of South Dakota, Idaho, Montana, and again, Wyoming. 6 percent have come from outside of the United States. In the earlier years, at least, the predominant geographical location was Massachusetts. And, it seems as though as Massachusetts was declining the rest of the USA, outside of New England was increasing. Also, it is interesting to note here that the foreign contribution has been quite substantial in recent years, between 20 percent and 30 percent. A similar distribution for the carotidal melanoma program. New England has contributed the most – at least in the earlier years. I notice that in the last three or four years, that the dominant contribution from Massachusetts and the New England States, outside of Massachusetts with the rest of the United States contributing just about the same percentage. And necroidal melanomas, also the foreign contribution has been relatively small. Less than 10 percent. If you one on top of the other, one thing stands out: in recent years, the contribution of Massachusetts is almost the same for both programs.

Concentrating just now on Massachusetts and the uveal melanomas. The accepted occurrence of choroidal melanoma is six patients per million population in the United States per year. We have saturated the patient population in Massachusetts. There have been a few outliers in 1992 and '95. But we seem to be swinging back, now the last three years to the magic number of 6. But for the neighboring States around Massachusetts, Maine, New Hampshire, Vermont, New York, Connecticut, and Rhode Island we have not saturated the patient market. These are again, just carotidal melanomas. And we've averaged, approximately, since 1983, 1.5 and 2 patients per million per year.

And, finally, these are all of our fractionated and uveal melanoma patients –
1974 through December of '98, 4,565. 4,079 came from the United States,
and they are all plotted here. Two things jump out at you. One is this rect-
angle right here, which has provided absolutely no patients. And that is the
State of Wyoming. Even though the Director of the Cyclotron was born and
grew up in Wyoming, he hasn't been able to attract anyone else to come this
way. He claims that nobody gets sick there! Wyoming has the least density
population of any State in the Union. Approximately 5 people per square
mile. That's one reason why. The other stand out is Massachusetts. It has provided most of our patients 1,365. No other State comes close. The closest
being New York, with approximately 400. As you might expect it is the more
populous States that have contributed the most patients, between 101, and
400. Those are on the Green. Florida, Illinois, Ohio, Pennsylvania, New York,
and most of the New England States, excluding Vermont. Vermont has contributed some 54. Finally, out of the total 4,564, 60% have come from within
a 500 mile radius of the Cyclotron. And 74% from within a 1,000 mile radius.

PAUL CHAPMAN

This talk is about history, and I will frame it in that context. Neurosurgical activity at the Harvard Cyclotron goes back almost as far as the fifty years
of HCL history we are celebrating today. The fabric of any history is really
woven out of the professional, and often personal, lives of the people who
were involved in those events. And the history of neurosurgery at HCL is
certainly a very colorful and rich tapestry. In that vein, I am going to mention
a number of people today in recounting this history. I apologize for leaving
anyone out, but will try to at least give you a sense of the individual contributions that have been of signal importance from a neurosurgical viewpoint.
You might wonder why neurosurgeons became involved with cyclotrons at
all. Sometimes history revolves about chance occurrences; however, if you
scrutinize the events closely enough, you can usually find causal antecedents.
The logical precedent in the case of proton radiosurgery was an intense interest within neurosurgery in developing precise methods for localizing targets
within the brain using mechanical three-dimensional coordinate systems and
x-ray control. These efforts made great progress and were successfully realized during the 1940s and 1950s. The method became known clinically as
stereotactic neurosurgery. The technique allows a surgeon to pass a probe
into the brain under x-ray control, and precisely make a lesion or stimulate
the brain in order to treat some abnormality such as a movement disorder.
    Target-centered devices are based on the principle that the target is al-
ways at the geometric center of the mechanical coordinate system. Regardless
of the position from which the probe is directed, which can be arbitrarily
chosen depending on mechanical constraints of the system or brain tissue to
be avoided in passing the probe, the same target point is precisely arrived at.
In 1951, Lars Leksell, a Swedish neurosurgeon, who has been one of the most
influential neurosurgeons of the twentieth century, coined the term radio-

surgery for using a single large dose of radiation, delivered stereotactically to a small target volume, to create a brain lesion. Interestingly, his initial work was with protons. He utilized the 185 MeV proton beam at Upsala for making functional lesions in patients with movement and mood disorders. As Andy Koehler indicated, Leksell and his colleagues couldn't reliably control the depth of penetration of the beam, i.e. the position of the Bragg peak. At that time, Leksell saw this as a method for curing brain lesions without reference to the usual radiobiologic principles of fractionation. This is what interested neurosurgeons. It interested neurosurgeons in Boston as much as it did those in Sweden. As mentioned by Professor Wilson and Andy Koehler, Dr. William Sweet, Chief of Neurosurgery at Massachusetts General Hospital at the time, was the person primarily responsible for applying the proton beam to clinical medicine at HCL. Dr. Sweet was really interested in boron-neutron capture therapy. But he had a mind like a trap for innovative and potentially useful ideas that could be applied to neurosurgical treatments. He called upon Raymond Kjellberg, a very bright, energetic fellow who had just completed his neurosurgical residency MGH, to assist with the project. Dr. Sweet assigned Ray the task of developing a technique for treating neurosurgical patients with the proton beam. And the rest is history.

Because so little was known of the effects of single dose or hypofractionated treatments, Dr. Kjellberg began by making lesions in monkey brains. He also combed the radiobiology literature for potentially useful material based on fractionated radiation treatments and experimentation. Adapting this information to his own ends was often intuitive or speculative. In preparing to treat patients with a single, large dose of Bragg peak radiation he was certainly entering poorly charted territory.

The first patient treated in 1961 was an unusual case, not the sort of patient that Kjellberg typically treated for so many years thereafter. She was a two-and-a-half-year old child with a malignant tumor of the central part of her brain. As you can see, she had already received a great deal of conventionally fractionated radiation. Other forms of treatment such as chemotherapy had been exhausted as well. And the tumor was still growing very rapidly. The treatment was performed under general anesthesia, which in itself was a remarkable feat at that time Because they really didn't have a firm idea how much radiation should be given, it was delivered in three fractions over several weeks. The total dose was 8,000 rads.

Two months after treatment there has been dramatic shrinkage of the lesion. Over succeeding months, the area that they had irradiated remained stable. But the tumor outside the treatment volume continued to grow and eventually took the child's life. At post-mortem examination, there was extensive necrosis without tumor in the area treated. The lethal part of the tumor was that which had grown outside the field of irradiation. As mentioned, a primary goal at that time was to establish what were safe and effective doses of Bragg peak radiation given as a single fraction. Dr. Kjellberg initially turned to two bodies of data to address this problem. The first was the clinical experience of radiotherapists, based on fractionated irradiations. Dr. Kjellberg used a so-called Strandquist curve, inserting data on whole brain radiations that had been gathered by Lindgren. and extrapolating the result back to a brief interval of radiation exposure. This required exercising some tentative assumptions about applying whole brain radiation parameters to the Strandquist curve. But, in fact, the results turned out to be quite useful. The second body of information that Kjellberg drew upon was the experimental monkey irradiations that Andy Koehler has described. Later, Kjellberg and his collaborators also studied the effects of the ongoing proton treatments that he was performing, principally involving patients with malignant tumors. This was done in conjunction with Andy Koehler and the Neuropathology Department at MGH. Using this material in a rather imaginative way, and combining it with some eclectic data from a variety of other sources within the radiobiology literature, Kjellberg constructed a family of curves relating risk of complication to dose and target size.

On this curve Kjellberg put points for two patients treated for glioblastomas. A second set of points depict Kjellberg's monkey irradiations using a 7 millimeter proton beam. A third set of data points is spinal cord irradiations that had been done many years before by Boden. Finally, these points represent the results of a series of microbeam irradiations performed by Zemann. Kjellberg applied a wide variety of disparate information to the problem of arriving at some predictable relationship between the size of the lesion being irradiated and the dose which could be safely used without causing radiation necrosis. Ultimately Kjellberg canonized his results into a set of curves, shown on page 42, which he continued to use throughout his career.

For most of his career, Kjellberg generally treated patients with a radiation dose that would have a predicted risk of radionecrosis of less than a 1 percent. The general validity of this graph as a close approximation of what is actually observed, is reflected by the fact that many other neurosurgeons who later began to treat patients with LINAC and gamma knife radiosurgery adopted his method of risk analysis and have continued to use it.

The ubiquitous water telescope that Andy so humbly described, which he and Bill Preston devised. We continue to use the water telescope up to the present time for the STAR radiosurgical treatments. There's a very important point to be made regarding the water telescope vis-a-vis radiosurgery: that is, it is absolutely essential when using a target-centered stereotactic device, such as Kjellberg's as well as our own STAR system. This is because one needs something that is going to maintain the Bragg peak at the same point in space, regardless of the orientation of the patient. This assures that the target to be irradiated, the Bragg peak, and point of convergence of the axes of rotation of the patient positioner all coincide as the patients position is changed. It is perhaps relevant that at the other two proton treatment centers in the United States, Berkeley and Loma Linda, target-centered treatment devices are not used, reflecting the fact that neurosurgeons provided little input during the developmental phases at those institutions.

In order to fully appreciate an account of Kjellberg's contributions, one must keep in mind that computer-based imaging such as CT and MRI scanning was not available for most of his early career. These technologies have since become available for diagnostic, and more recently, 3-D treatment planning purposes. We have come to rely heavily on them. They have also revolutionized stereotactic methodology. The first CT scanners, thanks to work that was done at HCL, only came into general use in the mid-1970s. Initially Kjellberg had to rely on skeletal landmarks visualized on skull x-rays to do his targetting. This is an important point, because it determined the types of lesions that he was able to treat. One category was tumors of the pituitary. This is because the sella turcica is easily localized on x-rays. The Berkeley group was also interested in treating pituitary lesions for similar reasons. By the same token, there was another not quite so obvious lesion which could also be visualized and projected on plain x-rays. And that was an arteriovenous malformation. For those of you who aren't familiar with AVMs, they are relatively common. And they are important clinically because they are a frequent cause of intracranial hemorrhage with resulting death or disability. They have an approximate 4–5% risk per year of hemorrhage. This risk is cumulative over a person's lifetime. With each hemorrhage there's about a 50 percent chance of the patient suffering either death or major permanent morbidity. AVMs consist of a tangle of thin-walled blood vessels that allow shunting of blood under high pressure from arteries directly into dilated veins. Prior to the advent of CT or MRI scanning, these lesions could be visualized by cerebral arteriography. By injecting a radioopaque liquid into the blood vessels supplying the brain. the material could be detected on plain skull x-rays as it circulated through the cerebral vessels. This allowed Kjellberg to use plain x-ray images to see cerebral blood vessels and the associated AVM. He was then able to precisely target the lesion, because its location could be referenced to bone landmarks of the skull, which were visibleon the same x-ray. At the time of the treatment, orthogonal x-rays of the skull were sufficient to localize the AVM with respect to the proton beam and Bragg peak.

Kjellberg treated the first AVM in 1965. He didn't treat the second until 1972. The 1965 treatment was the first time that stereotactic radiosurgery had been used for an AVM. In order to appreciate the landmark nature of that treatment, one must understand that, at the present time, stereotactic radiosurgery is a mainstay in the armamentarium for managing these difficult lesions throughout the world. Because there was no precedent for using radiosurgery for AVMs in 1965, Kjellberg most likely undertook the first treatment simply for lack of any other safe and effective therapy. It was 7 years before he treated another one. And, it was yet another 10 years before he had enough patients to publish his experience demonstrating the usefulness of this form of treatment. In the meantime, and even following his publication in the New England Journal of Medicine, there was considerable skepticism within the neurosurgical community. Ironically, resistance to the radiosurgical treatment of AVMs subsided rather dramatically in the 1980s when photon radiosurgical systems such as LINAC and gamma knife became available, allowing other neurosurgeons to carry out similar treatments on their own patients. In one example of an AVM treated by Dr. Kjellberg, the result 3 years after treatment shows that the lesion was gone (shown on page 40). Fifteen years ago, this result would have been considered miraculous. Now, we consider it an expected result.

Between 1961 and 1993, Kjellberg treated almost 3,000 patients with Bragg peak radiosurgery. Many of these patients had pituitary tumors. Most of these were hormonally active: that is, they secreted a substance such as growth hormone or another hormone which would cause a hyperfunctioning of the adrenals or, less commonly, the thyroid gland. He also treated some patients in order to suppress normal pituitary function, this being one of the earliest uses of the proton beam. By the mid-1980s. arteriovenous malformations comprised the lions share of Kjellberg's treatments. Ultimately, he treated more than 1300 AVMs. What do I consider Kjellberg's most important contributions? The first thing that I alluded to was that he established risk prediction standards for radiosurgery, based on dose and treatment volume. That work is cited up to the present day. His other major contribution was to demonstrate that one could effectively treat arteriovenous malformations with high-dose, single-fraction irradiation, whether it was protons or some other modality, such as x-rays or gamma rays. I recall attending a national neurosurgical meeting in the early 1980's at which Kjellberg presented his AVM data. He patiently weathered a storm of skeptical criticism concerning the ability to effectively treat arteriovenous malformations with proton radiosurgery. Within 10 years, people had applied the use gamma rays, the so-called gamma knife, and x-rays to the radiosurgical treatment of AVMs. In 1990, there were literally hundreds of facilities around the world actively involved in these treatments. By that time, three-dimensional computer based imaging had make it possible to treat a much wider range of lesions, particularly tumors.

Putting events into a historical context, by 1990 Kjellberg was ill, and his career was drawing to a close. Moreover, his methods had become quite dated vis-a-vis other radiosurgical techniques being used. These issues threatened to bring the radiosurgery program at HCL to a close, in spite of the inherent physical advantages of Bragg peak proton treatments. Nicholas Zervas, Chief of Neurosurgery at MGH, had the foresight and energy to begin to reorganize the radiosurgical program at HCL and bring it usefully into the future. Dr. Zervas interest in radiosurgery was rooted in his own earlier training, which was heavily involved with the discipline of stereotaxis. Early in his career he worked with Professor Talairach in Paris, who had developed the first clinical stereotactic apparatus utilizing an x-y-z coordinate system.

Dr. Zervas and I embarked on an effort to secure the future of the proton radiosurgery program and adapt it to the most recent technological innovations. The first thing we had to do was to bring the imaging methodology forward. As I mentioned, Kjellberg was still localizing lesions using skeletal landmarks seen on plain skull films. This was a serious anachronism. Other radiosurgery units, in particular those with LINAC or gamma knife radiation sources, were using sophisticated 3-D image-based treatment planning systems, which were already commercially available. We had to learn how to adapt such state-of-the-art technology to the radiosurgery effort at HCL. The second order of business was to design and fabricate a new patient-positioning device, which would allow us to take advantage of the newer stereotactic imaging technology. We would now be treating a wider range of lesions that could be visualized on CT and MRI. We were directed to Project Genesis, a design group in Cambridge, for help in our efforts. The result was the so-called STAR patient positioner, STAR being an acronym for Stereotactic Alignment for Radiosurgery. It is a large, target-centered, stereotactic frame, which suspends the patient entirely within it. The patient is supported in the supine position on a couch with the head immobilized by four point skeletal fixation of the skull. He or she is then precisely moved within an orthogonal x-y-z coordinate system to bring the intracranial target to the isocenter of the device. Once the patient is properly positioned, they can be rotated about either vertical and/or horizontal axes, keeping the target at the geometric center of the positioner. Space was a major constraint in determining the design. The treatment room is quite small, making it necessary to build the couch so that the patient could be lifted easily from a sitting to a supine position within a short distance at the foot of the positioner. At that time, the treatment room was in fact the same beam line as was also being used by Dr. Kjellberg. This required us to make the STAR device completely portable, so that it could be moved to and from a storage area in the hallway when it was not in use. This was accomplished by using pneumatic footpads of the sort that are found in industry for moving heavy equipment about a factory floor. In spite of the fact that STAR weighed as much as a small automobile, two operators could easily maneuver it in and out of the treatment room. The design engineers at Product Genesis did an excellent job; in fact, they won national and international engineering design awards for their work. Once STAR had been determined to be an engineering success, we needed to test its actual human factors before putting it to use.

One of the things that we had done in the context of retooling was to establish what has become an essential and fruitful collaboration with the MGH Department of Radiation Oncology for the first time This is Ethan Cascio, whom most of you know. This is John Purbrick, an employee of Product Genesis who designed the original control system. Over the next few years, as we worked the bugs out of the system, John continued to be a very well known and popular member of the team, showing up at a moments notice on any day of the week to solve a software or wiring problem. These two gentlemen, John Steichen and Bill Butler, were neurosurgical residents at that time. I brought them over to the Cyclotron to try to get one of them to volunteer to be the first patient in STAR but they both declined. That left me to do it. Bill Butler, with some trepidation, applied the stereotactic frame to my head, as shown here. He did such a good job that he's now on the Neurosurgical Staff at MGH. The sign, prepared by Ken Gall says Smile Paul, it won't hurt a bit. And it didn't. We were pleased with the results, and satisfied that patients could be treated with a reasonable degree of comfort.

Metastatic tumors are extremely responsive to radiosurgery. There is an efficacy of about 90 percent in either controlling or eliminating the tumor. With the improved technology of STAR, we are treating a wide variety of malignant and benign tumors such as inoperable pituitary adenomas, meningiomas, and acoustic neuromas. The last named is a benign tumor, arising from the auditory nerve at the base of the brain. We have arrived at the point where there is now serious debate within neurosurgery as to whether radiosurgery is preferable to an open operation, even when a tumor is operable. Acoustic neuromas are an example of this phenomenon. We continue to treat arteriovenous malformations, an important tradition begun by Dr. Kjellberg at HCL. One patient illustrates what makes the STAR radiosurgery program worthwhile. There was an inoperable AVM next to the brainstem in a 20-year-old girl. She was one of our early cases. She had suffered at least a dozen life-threatening hemorrhages, and was having a bleed every few months. Several years after treatment the large AVM had disappeared. She is also doing quite well clinically, without any further hemorrhages and leading a normal life. And, this is really what gives one a warm feeling about all of the effort that everybody has put into the STAR radiosurgery program.

Our collaboration with Dr. Suit and his colleagues has been most productive. Inparticular, I would like to mention Dr. Jay Loeffler, whose reputation throughout the world as a foremost expert on radiosurgery is largely responsible for a dramatic increase in the number of patients that we were now treating. Last month we treated our 500th patient, and we're now treating at the rate of 140 to 150 new patients a year. At this point we've reached a maximum, in terms of the capacity for patient treatments at HCL. So we're very much looking forward to moving over to the new cyclotron facility at MGH. However, today is really a time for looking back, not only to reminisce and take pride in our accomplishments, but more importantly to identify those things we have learned that should be carried into the future. Ray Kjellberg's material still represents one of the richest available sources of information about the effects of single-fraction radiation on the brain. His experience with more than 1,300 arteriovenous malformations is still the largest extant series. Sue Lyons was his nurse assistant for many years and knows many of these patients personally. With her indispensable help we have been able to contact and follow-up on over 90 percent of them. This has allowed us to critically refine his risk prediction paradigm in order to treat patients more safely and effectively.

We've identified which patients had complications in relation to the size of their AVMs and the radiation dose delivered. Using this, we generated much more accurate dose-response curves than were available to Kjellberg. The follow-up, accurate calculation of treatment volumes and doses, and the statistical treatment of the data has been an enormous effort, requiring the help of many individuals who are in the audience today. But the result is very important not only for our program but also for radiosurgery in general.

EVANGELOS GRAGOUDAS

It is a real pleasure to give you a review of the Eye Treatment Program. Certainly, it is something that changed my career, treating these patients over the last 25 years. I thought that I was the oldest one who uses modality. And this morning, I felt quite good. Because everybody was talking about experiences in the 1950s and 1960s. At that time I was in high school, and medical school. So, I was the youngest of the group and a great thing! The eye program started with a treatment of uveal melanoma. And, for the non-ophthalmologists, this is the most common primary interocular malignancy. It threatens sight and life.

The pigmented tumor is a tumor, a carotidal melanoma. When we started treating these patients back in 1975, almost all of these eyes had been inoculated. And you can see this eye that has been removed, because of a large tumor in the back part of the eye. A few patients were treated in Europe, using some radioactive cobalt sources. Later, those also were improved.

The list of the people that were involved in the Eye Treatment Program, especially at the beginning, includes a large number of ophthalmologists. Radiation oncologists. Bio-statisticians. Physicians. To me, the eye is the ideal organ in which to treat tumors. Because this is probably the only organ in which you can see the tumor very well. You can get pictures of it. You can measure it. And, in sub-millimeter precision. Certainly with new imaging techniques, as Dr. Chapman mentioned, the MRIs and the CAT-Scans, you can see tumors in the brain. But, nothing is so well outlined, as getting a front photograph of the eye where you can look into the tumor in great detail. The optic nerve is about 1.5 millimeters in diameter can be seen in millimeter precision. Protons can give you a uniform radiation dose and you can probably even localize the dose. We compared the dose distributions between the two modalities, cobalt and protons. You can give the whole eye a lot of radiation. You can radiate the macular with a substantial dose. The lens we can destroy with radiation. With protons you can give no radiation to the macular, and nothing to the optic nerve. And only a small part of the lens is irradiated.

We first produced photographs of a monkey that was treated by protons. He was did a lot of work in establishing this modality, because the whole eye program started in an unusual way. Dr. Charlie Regan, who was the predecessor of mine in the Retina Service, got the first grant to treat patients with abnornal blastoma But then he became the Acting Chairman of the Eye & Ear Infirmary and had no time. So he asked John Constable to get into the program. John, from Australia, treated a large number of monkeys. And he showed, the sharp demarcation of the beam. And, before he was leaving for Australia, he got me into the program. And I have continued since then. And, I follow these monkeys. You can see the radiated part, and the non-radiated part. And you can see the photo-receptors in all the areas of the retina, and they are normal. I did some electro-microscopy also. Half a milli-meter away from the radiated area the photo-receptors are normal. I did some digest preparation to look at the vessels. And you can see the radiated vessels, versus the non-radiated one. It was is a very highly localized

At the present time, there is a number of centers that now treat uveal melanomas with protons. It took some years, but now you can see a number of these facilities which keep increasing. In Berlin, Germany one recently started. They have treated over 50 patients. The way that we decided to treat this is to localize this tumor. We put some sutures through the muscles here, and we use a fiber optic light source here, to transluminate the eye. And, you can see how well you can see an outline of the tumor on the sclera on the white part of the eye. And then, we suture some small metallic rings from titanium which is an inert metal. And we transluminate again, and we can see the location of the rings relative to the tumor. So, we can adjust the aperture during the treatment. We use a computer program that Mike Goitein put together many years ago. You can see here the optic disk in the front of the eye. The macular here. You can see the tumor with the rings around it. And, you can manipulate the gaze in real time. So you can see exactly how much radiation you give to the different structures of the eye. You can see that the macular is outside of the beam. The same thing with the lens, the cornea, and the optic nerve. We have a video camera, and we can see the front part of the eye outside. And, we outline this on the screen. And, we take the final x-ray and this here has been established before from the Computer Planning Program. We can see the rings inside the aperture. And this is 20 times magnification of the front part of the eye. So, we can see a 1 centimeter movement on this screen, which is about half a millimeter in real life. At the beginning, we were thinking about electronic interlocks, in case that the patient moves, but when we tried it, it was quite obvious that we could sit down and look at a small object for minutes, without any problems. So, it is rather rare that we had to re-position the patient.

I give you an idea about the long-term results of treating these patients. Because, as you know, radiation can have complications many years after the treatment. And, at the present time, more than 6,000 or so were treated worldwide with protons. We ourselves, as Kristen Johnson mentioned, have treated more than 2,500 at the Harvard Cyclotron. The youngest we treated was 14 years. The majority of the patients are elderly, around 60 to 75. We treat very large tumors, because nowadays, although there are quite a lot of radiative plugs used like iodine and ruthinium, still the large tumors are not well treated with these modalities, and we get quite a lot of referrals from centers that use plaques. All the tumors that we've treated, show regression, the rate of regression. Although it is a lot, it is rather rare that we have complete disappearance of the tumor. Most of the tumors regress, but still something is still present. The remaining tumor can stay there for years and years. We believe that those are sterile tumors. And they have no ability not to proliferate or metastasize.

The kind of vision you expect from these patients depends on a lot of factors. It depends how far away it is from the macular which is the area of the central vision, or the optic nerve. As well as the size of the tumor and the initial visual acuity. Some of these patients have large retinal detachments. And, even if the retina goes back in place, the vision doesn't recover. But the results are quite good, considering the fact that we treat a very large number of patients with extremely large tumors. As you can see, about 40%, have a visual acuity of 20 to 100 or better. And that allows you to take care of yourself, although you have difficulty in reading. Although sometimes you can read with visual aids. But, about one-fifth of them have visual acuity of 25/50, or better, which allows you to drive a car, or to read. If you have a tumor at the periphery of the eye, away from the macular and the optic nerve, this is posttreatment. You are going to have regression of this tumor, and the visual acuity is going to be quite good. However, if the tumor is in the macular, in spite of the fact that this tumor is regressed it is not unusual to see complications – mainly vascular complications. The vessels of the eye show obstruction, accidents and things of that sort.

The local control of these tumors is phenomenal. 97 percent of the tumors that we treat are controlled locally. Only 3 percent they can show recurrence. And, most of those are marginal recurrences. The treated eye is retained in 90 percent of the cases. The eyes that end up being removed later, are mainly because of complications. And usually, this is neuro-vascular glaucoma, which is high pressure in the eye. And becomes a painful eye.

There is a whole list of complications. You can have radiation damage to the macular and the vessels that nourish the macular. Or to the optic nerve. And your vascular glaucoma, which is high pressure. Radiation cataract, we see quite a lot – depending on the dose that the lens is going to receive. This is not a major problem because the lens can be removed and can be replaced. Some of these complications are more difficult.

Obviously, any time that you deal with malignant tumors, metastasis is a major issue. And, from this series that I'm reviewing for you now, about 25 percent died from metastatic disease. And this is not different than the group of patients that have removed the eyes long time ago. The survival rates, about 83 percent at five years, and, as I said again, this is comparable with the patients that had a nucleation. There are several risk factors for metastatis. Obviously, the size of the tumor is the most important. The largest diameter was the prognosis. If the tumor is outside of the globe also, it is by prognosis. And, involvement of the ciliary body which is close to the anterior part of the eye, also is a key prognosis. And the older the patient, the worse the prognosis. You can see, for example, the difference between the different diameters. And, you can see if you have over 70 millimeters in diameters, the metastatic rate is much higher.

So, in summary, the Melanoma Program with the long-term results, show that the local control is excellent, about 91% of the patients. The tumor is controlled inside the eye. Many of the patients retain useful vision. And 75% survive without metastasis for 10 years.

Since we have such a great control inside the eye, we thought that probably the dose that we give is high and perhaps we could reduce the dose. We completed the randomized trial, between 50 and 70 Gray, and the recurrence rate is the same in the both groups. However, the visual acuity was not better. We still follow these patients for a long time, to find out if this is true. We have a protocol now. We give interferon therapy in high risk melanoma patients. The question of hyperbaric oxygen, and its usefulness is still under consideration. And, also we're working with photo-dynamic therapy to see if it can be used, in addition to proton beam radiation. This is not the only eye program, although this involves the majority of the patients. A few years back we started treating retinal blastoma, which was the original idea to begin with. Dr. Mukhai is in charge of this program.

We have expanded this program to treat metastatic tumors. This is palliative treatment. Most of these patients die. But, right now, we have quite a lot of patients that live four and five years, after the development of metastatic disease. And, obviously, you know, these patients need their eyes. And sometimes, you know, that's the only thing they do. They read, or they watch television. And, this is a patient with metastatic breast, for example. You can see a large tumor on the iris. And this is post-treatment. It's a very straightforward, easy treatment. We give only two fractions. We have treated some benign tumors also. Vascular tumors inside the eye.

One program that we have, that I will not discuss, because it's still under investigation, is the Macular Degeneration Program. Macular degeneration is one devastating disease. And it's the most common cause of blindness in the Western world. It's secondary to the development of new vessels in the macular, which is the area of the central vision. The precepts are, again, the same. That the highly proliferating vascular endothelial cells are going to be more sensitive to radiation, versus the retinal vascular cells. And at the present time, we are investigating this. We treated about 200 patients, right now. And we are in the process of doing some analysis. We use a two-fraction – two different kind of therapeutic schemes – to see if one is better than the other.

Our first patient Mr. William McKelvey from Colorado called us before we were ready to treat the first patient. We were planning to treat at least three or four months later. He had found out about it through his ophthalmologist who was a Resident of the Eye & Ear Infirmary. He said I apply to come and have the treatment of my melanoma. Nobody was ready. We didn't even have a chair for him to sit. And, he went all around the country. And, everybody was advising him to have the eye removed. He refused. And so, he came and it was a tremendous activity at the Eye & Ear, and at the General, and we had a Cyclotron. We measured the tumor. We took a lot pictures, left and right. And he is still alive and well in Colorado. We treated him in July 1975. His visual acuity is about 2200. And, he still has his eye, and he is still with us. It was a successful program, and I think we'll continue. And I think that the new Cyclotron is going to give us a new impetus to treat even more patients with different eye problems.

HERMAN SUIT

I'm very pleased to be able to participate in this tribute to the Harvard Cyclotron and its many physics and medical achievements over a quite long period. This has involved many institutions and disciplines. First of course is the Harvard Cyclotron Laboratory Group, the Harvard Department of Physics, the Harvard Medical School, the Massachusetts General Hospital, the Massachusetts Eye and Ear Infirmary, the National Cancer Institute and many others for their critical support, over quite a long time period in this area.

Ernest Rutherford, was responsible for our becoming aware of the protons, back in the early '20s at the Cavendish by his experiment of irradiating nitrogen gas by alpha particles and observing the product of O17 and protons. Actually this was the first alchemy, viz artificial transmutation of an element. He named them protons after the Greek word protos or first. Robert Wilson is the gentleman who wrote the paper in 1946, in which he proposed that protons would have an advantage in the radiation treatment of the cancer patient. He not only developed the rationale for proton radiation therapy but also, as Andy has mentioned, the rotating wheel of variable thickness to generate the appropriate distribution of proton energies in order to provide the desired spread out Bragg peak.

Now, the people to whom I would like to give credit include the Cyclotron Laboratory team, Professor Wilson, Bill Preston, Andy Koehler, Miles Wagner, Bernie Gottschalk, Janet Sisterson, and Kris Johnson. At the Harvard Medical School, – Dr. James Adelstein chaired with great effectiveness the Cyclotron Operating Committee. At the MGH, many excellent physicists have been essential to any success we have enjoyed. These include: Michael Goitein, Al Smith, Skip Rosenthal, Wayne Neuhauser, Hanne Kooy, George Chen and Judy Adams. Marcia Urie, Lynn Verhey and Ken Gall have moved to other medical centers. There has been an excellent team of physicians on this program. These include John Munzenreider (one of the longest serving participants), Norbert Leibisch, Mary Austen-Seymour, who has moved to the University of Washington, the late Ira Spiro, and Alan Thornton. New participants include Dr. Jay Loeffler, who is the Medical Director of the Program, Thomas Delaney, Nancy Tarbell and others. coming on-line.

I would like to comment briefly on why we were interested, and so absolutely keen to get into proton therapy. Treatment of tumor by radiation, using conventional modalities, still carried a worthwhile probability of failure to eradicate the disease. Additionally, there is an associated not insignificant risk of morbidity. Recall that unsuccessful treatment means death of the patient, unless there's a salvage treatment, which is effective, available, and could be applied. And, we know very well, that in many instances, salvage treatment is not effective, or it cannot be applied, or may not be available. Estimated risk of radiation morbidity is the principle determinant of dose that one can give. Complication almost invariably this arises in uninvolved normal tissue adjacent or tissues near the target. Complications due to the direct effects of radiation never arise in un-irradiated tissue. Thus, there is a clear advantage to reducing irradiation of non-target tissue. The use of the SOBP with a near uniform dose across the target volume with virtually no dose deep to the target was and is an obviously superior radiation beam for therapy. Reduction of the treatment volume, permits an increment in dose to the target and hence, and increase in tumor control probability. Further this lesser volume of irradiated normal tissues means decreased frequency and severity of complications.

I came to MGH in 1970 with a background in experimental radiation treatment and was, accordingly, very familiar with the dose response curves for tumor control. These studies consistently demonstrated for spontaneous and their early generation transplants in syngenic animals an orderly increase in response frequency with each increment in dose. These curves were rather steep, ie there would be a useful gain by raising dose by 5–10%. The dose response curve for human tumors was judged to be similar but slightly less steep. Accordingly, as the physical dose in proton therapy techniques would produce a smaller high dose volume, there would be tolerated a higher dose to the tumor and therefore an improved tumor control probability would be realized. Oddly enough, this was not something that was uniformly accepted by radiation oncologists of the day. The only questions in my mind were: how big is the gain and is what is the cost in time, effort and dollars.

I met Dick Wilson soon after coming to MGH. He introduced me to Michael Goitein, and, in 1972, Michael Goitein joined our group. Michael and I and the rest of the people organized a Symposium at the MGH in 1972 on the physics and radiation biology of proton beam therapy in 1972. Biological experiments commenced in 1972.We treated the first cancer patient by standard fractionated proton therapy in January 1974. We were awarded a NCI grant to support the clinical study of proton therapy for the cancer patient in '76. This grant has continued to the present time and has been a tremendously important support for the program.

The HCL has been a cordial and friendly place for us to work as partners in this effort. We have been at the HCL four days out of five. A question immediately came up as we were starting to employ protons for curative therapy by more than a few concerned persons. This was the risk of marginal failures due to use of too small of a treatment volume. I was questioned on this by a very prominent physicist at an international meeting. He said, don't you think that you're going to be too clever by half? You risk missing the tumor, and so you'll lose on one side, what you gain on the other. My answer was a clear no, as there had not been an increase in marginal failures as treatment volumes had been progressively decreased over the preceding 1–2 decades. Normal structures which are not suspected of involvement by tumor [eg bowel in the treatment of prostate cancer, spinal cord in treatment of lung cancer] can be defined, and excluded from the high dose volume with impunity. Since 1950, there have all been many major technological advances, each directed to reducing the treatment volume. These include: introduction of portal films, simulators, compensating filters, 360-degree gantries, high energy photon beams [cobalt, linear accelerators], computer-based treatment planning, electrons, protons, intra-operative electron beam, stereotactic radiosurgery. Recent gains in planning/delivery include conformal treatment planning, Cuts, MRIs, ultrasound, PET, on-line portal imaging gating, sophisticated immobilization. What are we expecting now? These are all en route: increased accuracy and precision of gantries, patient support assemblies, Monte Carlo dose calculations, optimization of planning to include the biological effect distribution, IMPT – Intensity Modulated Proton Therapy and vast increased quality of imaging techniques.

Our first patient was a small boy who had a sarcoma (rhabdomyosarcoma) in the posterior pelvis. I thought he presented an entirely appropriate tumor normal tissue combination and that we should employ proton beam technique so as to avoid dose to the central pelvic contents. Michael Goitein, Joel Tepper, a resident in radiation oncology [now Chief at UNC], Andy, Bob Schneider and the whole group worked to complete planning and then perform the actual treatment. He had virtually no GI symptoms. Unfortunately, he later developed metastatic disease and died.

John Munzenreider and Norbert Liebsch have treated a very large number of the rare malignant tumors of skull base, viz chordomas and chondrosarcomas. They have local control results at 10 years of 45% and 95%, for the two tumor types, respectively. These results are substantially better than those of conventional photon therapy. For the uveal melanomas, the results have been most excellent with local control results at 10 years of 95%. Michael developed the first 3 D treatment planning system and it was for the uveal melanomas.

Now, we have active Phase III trials in progress for skull based tumors, uveal melanoma, prostate cancer and meningiomas. The contributions from the Physics Group have been of very high value. I wish to mention several by Michael Goitein. The first 3-D Treatment Planning system put into practical clinical practice was that of Michael's for the Eye Program. This has been the basis for our being able to do the eye treatments, and do them accurately, and know exactly where we are putting the dose. Further, this program has been provided to a number of centers, around the world. Additionally it was the foundation of the later more sophisticated planning systems used in our program. He also originated the concept of dose volume histograms, digital reconstructed radiographs, beam's eye view among others and display of uncertainty bands around individual particular isodosecontours.

I wish also to give credit to the 24 fellows who have spent at least one year in the program. Many have continued a major interest in proton therapy. We are expecting to participate in international Phase I, II, and III trials. Hopefully, these will include collaboration with the C^{12} ion programs so as to obtain evidence of the relative merit of protons with RBE of 1.1 and the 12C beams with a complex and higher RBE structure. Clearly, the program will need the continued support of The National Cancer Institute. Dr. Jay Loeffler, is the Medical Director of the Program and is leading the establishment of the program at the NPTC on the campus of the MGH.

MICHAEL GOITEIN

It was mentioned by a few of the speakers this morning, how central it was to their experience of this technologic star here. Of the people and the personalities with whom they came in contact at work. And, as I look around this room, I am reminded of how fascinating a group you all are. I know 70% of you. I've enjoyed this professional career with you. Most unfortunately, the statement that has most resonated with this, is the comment attributed to Dick that we learn by our mistakes. And, I realize how very true that was, for me even my bitter mistakes, or personal mistakes. And, this leads me to make a little observation, which I learned when I joined the Faculty at Mass. General. There's one rather critical difference between pure physics, and an application that involves people. And, as a pure physicist, I made an awful lot of mistakes. In fact, my apparatus, as Dick will sometimes remind me – often didn't work. You don't have that luxury in treating patients. You have to be extremely correct, all the time. And this makes a very, very fundamental difference in the disciplines. And, it also leads me to mention that there's a big difference in an applied discipline, as compared to a pure research. And, the difference is, is that the Center of Gravity belongs to the Application. And, I think I do want to say, without in any way detracting from the enormous contributions of everybody within the technical team that have made this tool that the Cyclotron lab has been very fortunate in its clinical colleagues. And, throughout the world, as Janet will tell you afterwards, protons have been disseminated. And, this has been because the Medical/Scientific programs have been conducted with enormous integrity and focus and success. And, that's extremely important. Some of the things that you've been shown; those of you not medically knowledgeable may not realize what they really mean. To have treated over 600 chondromas, and chondral sarcomas is a miracle. Before this program came into being, even a large place like Mass. General only saw five patients a year of these diseases. It is a very rare tumor. And this number attests to the fact that patients come from all over the world for one reason only. Because this has been developed as a very successful treatment. Proton therapy has become an international activity, as you can be reminded. But, I've lived that out in my personal life too. I bring you greetings from the Paul Sherrer Institute. Happy Birthday from them. I dare not try it in German. A facility is being built at the Mass. General Hospital with a Cyclotron in several treatment rooms called the Northeast Proton Therapy Center. This development rests on the Harvard Cyclotron development. The Harvard Cyclotron lives in this machine. The Northeast Proton Therapy Center is a more compactly-shaped structure. It has another little problem too. Mass. General about 100 years ago was almost in the Charles River. Patients were brought up by boat, not ambulances, for treatment. The new facility is built on land-fill. It stands on 362, 100-foot long piles. It's a construction feat. The Cyclotron, The Harvard Cyclotron Lab is contrasted here with the new machine. The Harvard Cyclotron has something in excess of 700 tons of steel which as you've heard produces 160 MeV. NPTC has 200 tons of steel with

230 MeV. It's a technological feat, I believe, to have been able to produce a machine, a successful machine, this compact. And this lightweight. If 200 tons can be considered lightweight.

I do want to say that I have learned many lessons from Andy and the Cyclotron. But, none more powerfully than the concept that simple is beautiful. And I do not believe that the complexity that we've laid on in the new facility is any advantage. And, if we do anything, it will be to try to preserve the important knowledge in that concept.

I want to point out though, that good though it may be, a single beam of radiation, or of protons isn't a very good way of treating a patient. It is now typical to use beams coming from many different directions. These fields are often very non-uniform. This is a rather modern idea that, in order to get a uniform dose in the middle, you do not need uniform doses from each of the contributing beams. And this is an idea, actually, whose origins lay in some of the consequences, in the work that Allan Cormack did on inverse reconstruction for Computed Tomography. Modern x-ray technology produces very good distributions. The good news is that the protons produce even better ones. But the challenge is that we have to match this relatively complex technology.

In the new facility, we have beam which rotates about the patient. In the nozzle, is a whole bunch of stuff. This stuff is nothing more, nor less, than all the stuff that's in the Harvard Cyclotron, but some are further elaborated. We are moving to, but are not yet at a rather different concept of sweeping the beam through the patient. So, instead of scattering the beam, we use a pair of magnets to scan the beam through the patient. And this idea too was fore-shadowed by a development that Andy had, in which he had a rotating magnet, to spread out the beam, and build a little prototype which, I think, is still hanging around the cyclotron.

There is a quite different kind of technology with a compact scanner, which is being developed abroad. The whole goal behind the new facility is for us to take and apply the techniques which were developed at the Cyclotron Lab. And to go beyond them. To take the medical programs that have been so successful there, and extend them using a machine that has more energy, so can reach any place in the body. And much more capacity, so that we can treat many more patients. And, we hope, with this to move into the future with a strong and very vigorous program.

JANET SISTERSON

I want to begin by talking about proton therapy in the world. And then, I am going to return and talk about physics research projects that have been ongoing at HCL. Some for medical applications, and some that are not. And, also to mention our outside user program, which some of you may be familiar with. And, again, others may not. If we look at proton therapy in the world, at the moment there are 17 operating proton therapy centers. There are many more who will come on line. And, when I went through my own mind, and I talked to Bernie, we discovered that we have helped design the beam lines at

least at six of these centers over the years. Proton therapy. We haven't saturated the need, or the requirement for the machines. If I take a very conservative viewpoint, of the machines that I think will actually come into existence
and operation the next few years, we can be up to at least 30 by the early
2000s. If we look at the patient statistics, worldwide, and these are cumulative
patient totals. Again, we have a log-scale. And that is dear to Andy's heart.
And I'm showing when the various patient treatments started. And I have
broken this down into the total patients. The Black is the world. North America
is represented by the red line. The Russian, the European experience is the
green data. And, it has a flat area in the '70s because some of the original work
was done in Europe at Upsala. But, really, the number of centers that opened
up didn't start until about the 1980's. The Russians who have contributed
much to proton therapy over the years and are still doing so. But we are all
aware of their economic problems, which prevents them, perhaps, from treating as many patients that they would like. Although I did hear on Friday that
Dubna, who has not treated patients for about the last three years, has now
returned to the fold of Operating Centers, and has treated a patient this year.
The Japanese too, have a significant contribution. And our only Southern
hemisphere, a proton therapy facility in South Africa, is certainly coming along
at a good pace. I collect these statistics, in part, because of my efforts with
doing the newsletter, which I'll mention in a minute for Proton Therapy. And
the last year that I have complete data in 1997. And this represents a 12-month
period that people will report their data. It may not necessarily be a calendar
year, but represents a 12-month total. And, at that time, there were 14 centers
operating. And over 2,000 patients were treated in that year. It's interesting to
note that Harvard is still treating about 20 percent of the patients, worldwide. Loma Linda, which is the only hospital-based facility, and has two gantries, three gantries, and can treat many more patients than we can. And, it is
obviously now treating most patients in the world. But, I would also like to
point out that this isn't a pie chart dominated by just two slices. It's a pie chart
where there are many facilities providing significant numbers of patients. And
what is hidden in this kind of pie chart is that some of them do that, when
they only get beam time for a very few weeks in the year. For instance, at
Upsala, which is one of the smaller segments up here, they treat about 50
patients a year. But I don't know how many weeks that's in there. That might
be four or five. The same is true at some of the Japanese facilities. The Boston
group, which must include both Mass. General and the Harvard Cyclotron,
have been very active in the worldwide arena. That has already been alluded
to. And, one of the things that we have, is a proton therapy cooperative group.
It's a very loose organization that really exists as a mailing list on my computer. And, Michael Goitein is chairman, and I am secretary. And, I also edit
this newsletter, Particles, which I put out twice a year. And it is through those
efforts that I've been able to maintain these statistics, as I try and collect them,
and I keep updated patient totals.

Every six months I try and update the patient numbers. I also try and keep track of who, what, why and when might be going to open a new facility. Through December, 23,000 people have been treated with proton beams worldwide. And, about over 27,000 have been treated with all of the heavy particles, which include helium, carbon or neon beams.

I would now like to return back to physics research at HCL. And I've got three segments to this. I'm going to briefly mention some of the HCL projects that we have done over the years since the end of the ONR program.. We've already heard about some of them. I would also like to talk about what we call "The Outside User Program". If you've got money and application, you can buy time at Harvard. And, very many interesting things are done there. And then, a program that in fact, I have been running for a long time, which is supported by NASA, which I have described here as "cosmic ray studies". So, just to highlight over the years. Dick Wilson mentioned Richard Albert, and our calcium work. He was our first graduate student on medical work. And got his thesis on this. And, we then extended that slightly, a little later on, to try and see if we could measure the calcium phosphorous metal ratio in bone. And, theoretically, you can. Practically, in bone, we really couldn't do it. But, along the way we measured some more cross-sections. We also had Bob Schneider who, I think in conjunction, well, I think it was Brigham & Womens', had been looking at producing rubidium that was going to be used in heart studies. And this was work that we initiated. And, another large component that has been gone on for, I think, the entire time at the Cyclotron. And, which culminated in Bernie Gottschalk's paper in the early 1990s. Is we've really done an extensive study for multiple coulomb scattering. We've made a lot of measurements. I believe they started in the 1960s with Andy. And, Bernie put it altogether, and did the theoretical analysis.

I'll now go to the outside users. I just found out, in fact, yesterday when I was putting this together, that the first outside user (Walter Brown) was a freebie in 1961. Andy recalled that he was a former graduate of Bob Pound's who was working at Bell Telephone Labs. If we now go to what is the current status, we only have time available for people to purchase time at the weekends. That's obvious. You've heard what we're doing all the rest of the time. And, it is of interest to note that our current income, from our outside user program and it is $350 dollars an hour. I believe it's the current rate… is about 10 percent of the total income of the Cyclotron laboratory. And, we are planning on moving this program to NTTC where the slightly higher energy and the continuous beam will be of interest to many of the people who come to us. So, why do people come to us? Often for cosmic ray studies. I am going to use an example which is the work of Jim Zeigler from IBM who has been kind enough to share some overheads, and information with me, to put together this small segment of my talk. And, he's a long-term outside user of the Harvard Cyclotron. He's been coming for about 15 years. He has studied "soft fails" in computer chips. But I also added in lots of other things, because he's quite an expert on terrestrial cosmic rays. And, he's in a terrestrial environment. And, if we look at a cosmic ray, when it comes to Earth, it will interact with the atmosphere. And, you really get a cosmic ray shower. So you don't get primary cosmic ray interactions actually on earth. You get the results of the showers which are, I believe, mainly protons, neutrons and some pions. And what they have primarily been studying is the soft fail rate in computers. And, there are two mnemonics used for this depending whether you're in the industrial arena, or the military arena.

I will continue to use the word "soft-fails", and not get into the politics involved. Cosmic rays will cause information in a computer to change. And, two things I hadn't realized until I was reading a paper of Jim's is this leads really to cautious users. You spend double the time, trying to double-check everything, so that you have the necessary redundancy. Another fact that I had never thought of is that if you have a mainframe computer under a service contract which stops functioning, the "soft-fail", you tend to replace parts. But it may be that the part did not need to be replaced. And, schematically, this is all just to give you an indication of what goes on. And, so you have a cosmic ray that comes in. And, the cosmic ray does a nice splat. It has a nuclear interaction. This may be enough to upset most of the electronic logic states. If you had a "1" it would now become a "0" or whatever. This problem has got worse, as our sophistication and the methods of building computers gets better. Because when the memories were very small everything took up a lot of space. Then, you didn't have as many of these fail rates. But, as soon as you started to make the much more complex chips, you can well imagine that it is going to get worse because everything has to be packed together. What I also found out is that really, things began to change in the 1990s. They got a much denser chip design in 1992. Went up to the next step. And now, the problem was that designers were running out of space. And they had to find ways to decrease the size of all of their components. They found three distinct solutions. Jim did experiments… radiation damage studies. They tested 26 different kinds of chip designs, all of which were basically electronically equivalent though they would be physically different. And, it's dramatic design here. And the showed that the cosmic soft error rate on fails, is very different, depending on the design of the chip. The fail rates, depending on the design that you chose for these components could be as much as one or two months. Or it could be better; one in two years. That's better. Or, if you really use the best design, it might be one per a thousand years. You know what you would do. You don't want your computer to fail. You would use this last one. And, another interesting point to note is that if you're at airplane altitudes everything gets worse, because there are more cosmic rays up there.

So, for this guy up here, the fail rate of one per two months for this design apparently translates into about one fail for every flight! So, the conclusion is that the fail rates are directly associated with both a chip design, and the way that you make everything. And, Jim also shared with me this noteworthy comment: two of the major pacemaker companies have reported significant problems with pacemaker lock-ups during airplane flights. And, now, these designs are tested at HCL by Jim Zeigler, when he comes to do his runs. And, I have another little anecdote on that too. Because I attend the Lunar and Planetary Sites Meetings. And two years ago, a colleague from Los Alamos introduced me to an astronaut in training, and said, Janet will be able to answer your question. Well, I couldn't. But I knew enough to tell him to ask Jim Zeigler. And the question was: why did all of our PCs fail when we went up to fix the Hubble? Because at one time, they apparently all just kicked the bucket and turned over and died. And, when I shared this with Jim, he laughed and said, of course they would. They weren't designed to be used in that radiation environment. So, I think it illustrates that by coming to Harvard, or somewhere like that you can learn a lot. You can use a proton accelerator to simulate stuff that you will find out either in space, or in the terrestrial environment. And, I think that we have seen, recently, an increase in uses coming from industry who have set up their own standards for equipment and designs. Or who are now aware of the fact that you can get these events happening. And they have to figure out how to design stuff better.

I'd now like to go to the extra-terrestrial environment, and cosmic rays. And here, we are concerned about primary cosmic ray interactions. And this is work that I have been doing quite a long time now. And, we first started some of this without any funding. But I've been funded by NASA since 1995. And my funding runs out in 2001. And, I am in the Cosmo-Chemistry Program in the Office of Space Sciences. And you might ask why? Maybe it will become apparent. And, the measurements that I make are to measure cross-sections that are needed to interpret the production of radio-nuclides and stable isotopes in extra-terrestrial materials. It sounds very easy. But it requires a multi-disciplinary, international collaboration. And, I would loosely call it a "cast of thousands". I added up the number of names in my last paper, and it was a lot. It was getting as bad as in high energy physics! If go out into space we have cosmic rays all around. I've mentioned an application on Earth where the primary cosmic rays actually hit the atmosphere and it is the showers that cause the trouble. But, out in space, where there is no atmosphere, on the Moon or in meteorites cosmic dust, or, all these other things in spacecraft, then the primary cosmic rays, actually interact with the object in question. And, very many things are the result of this.

I am particularly involved in two. You produce long lived radio-nuclides. Everyone knows about carbon-14 because it is used for dating here on Earth. But, it's an important component because of its 5,000-year half-life out in space. And, we also make stable nuclides, like neon isotopes, tritium and krypton. It is these isotopes that allow you to tell where the meteorites come from. And the Martian meteorite with life on Mars; it is that signature that enables them to know that rock came from that planet. And, what we can learn from that? Again, I'm more interested in these.

You can learn about the history of the cosmic rays themselves. And, whether they have varied in the past. We've only been able to make direct measurements of cosmic ray fluxes over the last few solar cycles since the early 1970's. By choosing appropriate radio-nuclides, you can look back over the past million years and see if we have had any sort of drastic events in those times. But, not only can we do that, but if we can also learn about the history of the object itself, and, for some of the lunar meteorites that found in Antarctica I've heard very beautiful presentations where they can figure out how long it has been in the Antarctic ice.

We have little pieces at different depths within this rock. The orange bits were sent to Tim Gerrold to measure. He measured carbon-14, which is his speciality. And so we can get a depth profile for the production of a nucleide. Another colleague of mine. Kernina Chozini from the University of California in Berkeley measured three isotopes in meteorite 64455, beryllium-10, chlorine-36, and aluminum-26. They all have different half lives, which embarrassingly, I can't remember what they are now. But the important feature here is that these are the measurements. They are different for beryllium from these two. And, if this is a depth and grams per centimeter squared, better density of a lunar rock somewhere between 2 and 3. So, if we divided by 2, this is about 1 centimeter depth. And that would be about 3 centimeters. So, anything down here is produced by galactic cosmic rays. And this increase near the surface is from solar cosmic rays. Beryllium is not produced very much by solar cosmic rays. It seems to be purely a galactic produced radionuclide.

There is only one physics equation in all of this, which is that really putting into an equation, the amount, let's say, of Carbon^{14} produced at a depth in a rock, as a function of a the production in all the elements that are in the rock? What is the cross-section, or the probability for producing Carbon-14 off that element? By all of the combinations of cosmic rays that we could find at that depth? The most important is component is the protons, because so many of the primary particles are protons. The next most important are actually neutrons at a depth. So, we need neutron production cross–sections. Until very recently, no neutron cross-sections were known. And, so, there was a lot of guesswork going on, and when the results weren't what one wanted, everybody said, there are no cross-sections, they must be wrong. In 1984, I wanted to ignore the theoretical lines, but to look at the few points that exist on this curve. And, particularly, for the cross-section of oxygen, of protons, incident and oxygen to produce beryllium-10. And these are the circles. And, this was all the data that was known worldwide in 1984. From few points we had to a complicated analysis. It's no wonder that people doubted the data. Particularly because the reason that they were doing it, it didn't seem to agree with what they thought it should.

If you look at the situation now, in 1996, and this is the work of another colleague, Bob Reidy from Los Alamos – the situation is very different. Although the graphs have points marked SI-96, which stands for Sisterson-'96 they are the results of the collaboration that I mentioned earlier, with all my cast of thousands. I'm not alone in this. And I've probably measured somewhere between 500 and 1,000 cross-sections using a very simple technique that requires very, very few corrections to the data. My colleague Ralph Michel from the University of Hanover has measured thousands of cross sections. But with a very complicated technique, with a lot of corrections required. And, the beauty of it is, when we agree, because we've used two different techniques we're pretty certain our results are right. There is still a dispute about a few points. And, in fact, I have repeated the measurements of those. We took them last year, and they're out being analyzed at the moment. This is a summary of the work we have done under this program. We've measured a lot of data. A lot of cross-sections. This was work started in the 1950's here at Harvard. And I'm always going back to the old data that was done here at Harvard. So do other people from outside. And we did these measurements at Harvard Cyclotron. And, it's very important to continue with them, because they are very basic building blocks if you'd like for many applications.

I remind you about cosmic rays. There are two kinds of cosmic rays. You have those that come from the sun, where nearly all of them are protons. 98 percent. And they are nowadays called solar protons, as opposed to solar cosmic rays. And, they generally have energies that are below about 200 MeV. And I'll remind everybody that those are numbers that we know here very well. NPTC will have an energy of about 230 MeV. Harvard Cyclotron has 160. So, we know already that the protons aren't going to penetrate very far into the surface of an object, only about two centimeters. We also know that the interactions that they will have will be fairly simple. And, that they won't make very many neutrons or other particles. And, we also know that these interactions are relatively rare. And most of them are going to be stopped by ionization, before they make the interactions. If, on the other hand, we look at Galactic Cosmic Rays, which come up from out in space, there aren't very many of them. And, they have a low flux.

And, I need to back-track one minute, because I did forget to say that at times of very great solar activity the flux of solar protons can be extremely high. They can be as high as three times 10^{10}. And I am going to get my units wrong. I think particles per centimeter squared per second. And that was a number that I really realized one day, corresponded to the Harvard Cyclotron, which is three times 10^{10} protons per second. That translates into about five nano-amps which is significant. You are going to get a therapy dose or worse. And, these bursts of radiation can last hours or days. There was a very big one in 1972. And there's been some very big ones recently. Galactic cosmic rays are also mainly protons. But they are very high energies, or can be very high energies. They can penetrate very deeply into the moon's surface, for example, or a meteorites. They generally have very big interactions. You get a lot of neutrons produced. And, what is also important is that those neutrons themselves, at depth and an object, then go on to interact. And, most of them interact when they don't just lose ionization. How do we have any objects to study? I mean this stuff is up there, and we're down here. Meteorites fall to Earth. There are a lot here. And, they can only be directly distinguished from terrestrial rocks if you really see them fall. Which are then known as a "Fall". Or you can find them. Some of them do look distinctive. Or, if you're in a place like Antarctica, they tend to collect on this blue ice. And appear as the ice moves up where there are some mountains or actually in the surface. So they have a whole team of people who go out every year, looking for meteorites. It reminds me of bird watching in Mt. Auburn, where you say, here's a bird, and everybody rushes around to look at the tree. They look at the meteorite, collect it and then, they try to establish whether it's an individual one, or whether it's come from a shard.

Alan Hills made famous in the book: *The Life in Mars* is one of the ones that was found in Antarctica. And, of course, in the movie of lunar missions. And I rather liked this. I can almost see him bouncing across the moon. And he has a little scoop and a rake. And they collected a lot of material to bring back. And it is from that material that we have been able to study radio-nuclide production in lunar rocks.

I show meteorite 68815, which in fact, I have been involved with. I have never seen it. But I know what has been measured in it by my colleague Tim Gerrold down in the University of Arizona, as you get a sliver of this cut out. You could tell which elements that lunar rocks are usually made of. And, these are the most principal ones. Oxygen is the most important element. Silicon is next. Others are trace elements. And you can get some others as well. The oxygen cross sections that were needed, had already been measured for radiation therapy. We had measured them because they were interference for the Cascio measurement. But, they represent some of the few measurements that have been made anywhere. Mark Chadwick at Los Alamos was very happy to have them, because he didn't know of any other measurements.

This work is ongoing. We are now extending it to do neutron production cross-section measurements. Of course, it is not easy to do that at Harvard, or at NPTC, although we did actually debate resurrecting the old neutron beam line that was in existence in the 1960's. But the beam intensity would really have not been high enough for us to do anything very easily. Therefore I have spent most of April, at the National Accelerator Center in South Africa making measurements, and also Los Alamos, last September. And, by doing all of that, we can then get, in summary, one of the things that we've been doing is that we have been able to learn much more about the variation of solar proton fluxes in the past. We need to know this, because the space missions are going out into environments for longer and longer. And, the hazards there is to both people, and to equipment. You need to design your spacecraft so that a person, or whatever, can get away from a really, really big flare. It is very difficult apparently to forecast these. They're trying to do that. It's also that some years ago, that people thought that within huge solar particle events, were affecting life on Earth. We still have a paleontologist who thinks that. And, in fact, I had extensive e-mails from him last week. Where, yet again, he's found lots of tracks in something – and thinks that we must have had a lot of protons falling on Earth. We don't think so, but he does. And, we need to do this, because we haven't been able to measure these for very long directly. Four solar cycles is very few over the existence of the sun, the Moon, and the

Earth. And, it kind of sort of blows my mind, you know, that from a little rock, or a meteorite that we can learn all of these things. These have all been ongoing. And, some of them are slightly different from the medical applications at the Cyclotron. But, on the other hand, they're also very similar. They are all involved solely with protons, or whatever – interacting with something. And, it's the same as with people. The nuclear interactions in the course of radiation therapy is a small component of what's going on. But I realized the other day that the neutron cross-section measurements that we are beginning to make, because of the materials I've included, will be very useful in de-commissioning studies of the Cyclotron, because we are – not only myself, but the Japanese will be making measurements. And you know have neutron cross-sections for of copper, nickel, and titanium. So, these applications are s all so far apart, and yet, they are also the same. So, I hope I've given you some flavor for some of the other things that have gone on at the same time over the last little while.

JAY LOEFFLER
        I have the dubious distinction of speaking at 5:30 after eight and a half hours of discussions. I'm sure that I should pass out some Ritalin so that people can have some attention. When I saw the title that Andy asked me to speak about, my first impression was, it's very easy to describe how we can make this program grow. And that would be for Michael Goitein to open up the new facility. Because, as soon as that happens, we will grow quite rapidly. I will talk about some of the things that we hope to do in the future. I think just very briefly, and I will not speak for 15 minutes, I promise. Talk about what have been the major accomplishments of this group. And I haven't really been part of this group. So, I accept no credit for any of these accomplishments. Despite the relative technical limitations of the Harvard Cyclotron, there's been tremendous clinical gain in certain disease sites that we have been able to treat. And, if you look at these funny disease sites, you saw over 2,500 patients have been treated with either uveal melanoma; over 600 patients have been treated for skull-based sarcomas. Almost 100 patients now have been treated with very unusual peri-nasal sinus tumors. Why these sites were picked is not because we had a particular interest in these sites, but these were sites that were difficult to control with standard surgery, standard conventional radio-therapy. And sites, we were technically able to treat. So we could escalate the dose, and achieve local control in the majority of the patients treated. We have an incredible group that's been put together by Dr. Suit, and Dr. Munzenreider and Leipsich. A problem was that Michael talked about physical research and development. We've trained people who have gone on to be leaders and leaders not only in our country. Markus Fitzek is here, who is now a leader in Berlin, in Germany Others are in Heidelburg, and France. And, we have a very strong group in biostatistics that's growing. And, a very strong record in publications both in the fractionated proton program, as well as in the radio-surgery program. What are our immediate goals? I haven't

been involved in this program for very long. It's amazing to me what great care is provided at the Harvard Cyclotron. I've had so many patients come up to me that I haven't treated, but have been treated by other colleagues in the department, and say, what a cozy, warm, comfortable environment the Harvard Cyclotron is. It's going to be very challenging for us, in this new facility treating four to five times the number of patients per day to maintain this patient-friendly environment. That's going to be a very difficult challenge. We also want to make protons part of the major scope and the armentarium of cancer treatment within the new partner structure which includes the Brigham & Womens' Hospital and the Dana Farber Cancer Institute. As well as, hopefully, collaborations with the Jimmy Fund, and Children's Hospital. Not just locally but with programs that we hope to expand with other proton facilities in the United States, and outside the United States. So the protons are considered this sort of niche field that is done at the Harvard Cyclotron, but part of the major cancer program within this healthcare system. It is our major interest, and our funding from the NCI is in this clinical research. And we are certainly going to expand this treatment program with the increased capabilities of the new cyclotron, and we hope to continue, as you heard from Kris, via the national, regional and international referral center.

The new themes for the clinical trials, pediatric malignancies, which we've treated some children with skull-based tumors, and a few patients with retinal blastoma… this is a major new goal and theme for our program. We also will for the first time, treating common malignancies. We treat some patients with prostate cancer on a randomization right now, between two dose levels early prostate patients. But we are going to expand that dramatically. We have programs in tumors of the gastro-intestinal system, including liver tumors. And, a program in the treatment of advanced non-small cell carcinoma lung. Part of cancer treatment in 1999 is not just using radiation alone. Most patients with cellular cancer are treated with combined modality therapy. What I mean by combined modality therapy is chemotherapy and radio-therapy. Plus or minus surgery. We have to understand the effects. By escalating the dose, and by reducing the amount of the radiation and the non target tissue what's going to happen to our ability to give combined modality therapy? Michael has postulated that we will be able to treat patients in combined modality therapy without giving treatment breaks, which are detrimental to the patient outcome, by reducing the amount of normal tissue irradiated. I think that there's very good reason to believe that will be possible. We are going to do Phase I trials – Dose Escalation Trials – which are very difficult to do with radiation. Because often, normal tissue outcomes take time. But we have a program that I'll talk about very briefly, for advanced lung cancer. Quality of life studies. If you can understand cost utility. Cost effectiveness. You have to understand what the true impact of your therapy is on patients' quality of life. And we are going to look at, in one study, the best of intensity modulated x-ray therapy that Drs. Suit and Goitein talked about briefly, the best of proton therapy.

I'll start off with Pediatric tumors. You think about a developing child, and the risk of irradiating a developing child. It's more than just possible limitations on growth and organ function. But also the potential of the radiation being associated with a second tumor. If you reduce the amount of the irradiated volume, you should reduce the incidence of second tumors. You should not affect growth and organ function as much. Some pediatric tumors are very curable right now. Early stage retinal blastoma is very curable with radiation. But, at a significant cost. And I'll talk about that, and how we plan to set up technology to avoid that. And we hope for other pediatric tumors where we need a higher dose of radiation. And, we'll be able to safely deliver this.

There are several disease sites that we have clinical protocols for. And we are asking for support from the National Cancer Institute – a tumor called medullar-blastoma. For the non-physicians in the audience, it's a seeding tumor of the central nervous system. Develops in the cerebellum. It gets into the cerebral spinal fluid, so that the whole brain, and the entire spinal access is at risk. And, they need to be irradiated in order to achieve curability in these children. Retinal blastoma and soft tissue and bone sarcomas in children.

I will mention the Archambault sisters. These are two girls that were born with the genetic type of retinal blastoma. They were cured with radio-therapy at Childrens' Hospital, back in the late 1970's, but at a cost. What is the cost of treatment? Well, you see the functional and cosmetic result of using x-ray therapy. You see the bony and soft tissue developmental changes in these children. But actually they can hide this quite well, if they change their hair. But that's just cosmetic. What's the more dangerous risk? Nancy Tarbell and colleagues were involved in a study published in The Journal of the American Medical Association. They looked at the actual risk of developing second tumors after treatment with radiation for retinal blastoma. And, these patients are genetically at risk for developing second tumors. But, if you add radiation therapy to this, you can see by 20 to 30 years very substantial risk of developing a second tumor from the radiation. And, the induction of second tumors are often fatal malignancies in these children. If you can reduce the amount of tissue irradiated can you continue to have high curability and reduce the incidence not only of cosmetic and functional effects but also of second tumors.

This is actually a simple plan for treating retinal blastoma with an anterior field that we can do as soon as we (NPTC) are open, using the gantry. Judy Adams and Al Smith were involved in designing this plan. We were actually able, in this plan, to reduce the maximum dose to the orbit to about 5 Gray. In the study, in the Journal of the American Medical Association that was basically the threshold for the induction of second tumors. So, we expect this to be a major success. We expect to have a large number of children in our country, referred to the NPTC for treatment for retinal blastoma.

Mesial blastoma. Is a disease in which the entire brain and spine are at risk. So, the entire brain and spine have to be irradiated. When I first talked to people about the possibility of treating patients with mesial blastoma, a seeding tumor, with radiation the whole brain spine needed to be treated. What would protons offer? It turns out that these children do get treated with combined modality therapy. Chemotherapy and radiation. And, one of the most active chemo-therapeutic drugs has an effect on hearing. It affects the cochlear nerve. And, it turns out with just the whole brain radiation alone, using protons versus compensated x-rays, you can reduce the dose to the cochlear nerve. Thereby hopefully, reducing the 60 to 80 percent incidence of auto-toxicity to this children, by using protons for the whole brain. And, more importantly, treating the spine. This is the common treatment worldwide now, a P-A field. And, you can see a substantial dose, to the lung, the media steinum, and the heart, using x-rays. With protons, as you know, you can stop the beam. And, we hope to do large animal studies to see what would be the effect if we don't treat the whole of the body; the point here is that you treat part of it. We affect growth, in a way. These children will develop scoliosis or kyphosis. Is it possible, perhaps, just to treat what the true target is? Avoid treating the entire of the body. Well, that is a little bit of a stretch right now. We'd be very experimental to do that. So we hope to do work on large animals first, to see what the effect of treating just the spinal chord would be.

It is very difficult to do a trial on photons versus protons. It's very difficult to do a trial here, where patients walk into the MGH, and they're randomized between x-rays and protons. Why do patients come to the Harvard Cyclotron? Why will they come to the NPTC? Well, they don't want to come for x-rays. They can stay home and get that. But one way of doing a comparison trial, is to do collaborative trials through National Research Programs. And, we're negotiating with a radiation therapy oncology group to do that. But, another way of doing it, is to do two simultaneous Phase I trials. And, this is for advanced non-small cell lung cancer. Use intensity modulated x-rays, and escalate the dose until we reach the maximum tolerated dose, and do a parallel study within intensity modulated protons. And when we reach the NTD of both, then, do a randomized trial. The disease, unfortunately, is common enough. We probably can do this in-house.

The other issues that I've talked about briefly, is the whole issue of combined modality therapy. We have a study that Chris Willard is running in our institution, once we open. You can randomize patients who need post-operative radio-therapy for rectal cancer. And they also receive concurrent chemotherapy to protons versus x-rays. The same dose. We're not asking a dose issue. We are asking: can you reduce the toxicity? Can you reduce the treatment breaks which are associated with a higher rate of pelvic failure?

And, finally, quality of life. Quality of life studies are mandatory in any cancer research program at the end of this century and all through the next. They're very expensive to do. We were very fortunate to bring in Jim Talcott, who is an expert in quality of life analysis particularly in patients receiving radio-therapy for prostate cancer. Do patient based quality of life assessment questionnaires. And this requires a new data base system. And requires us to bring in additional staff to do this. We eventually will be required, probably to do quality of life analyses for all of our clinical trials. And, based on what the quality of life is – and, based on the results and survival and local control and toxicity – we will be able to answer a very important question about the cost and utility of proton therapy in the treatment of cancer.

PUBLISHED WORK FROM HARVARD CYCLOTRON LABORATORY

This list is of all papers (that the webmaster can locate) of research that used either cyclotron by scientists from Harvard or other institutions, or papers by cyclotron staff. It does not include papers on work on other aspects of the Harvard-ONR nuclear physics contract such as Professor Ramsey's molecular beam work, Professor Pound's work on graviational red shift and so forth. Most internal laboratory reports are excluded.

1935–1945
"Some Effects of Rare Gases on Metal Spectra," William Preston, Ph.D. thesis (1936).

"The Focusing of Charged Particles by a Spherical Condenser," Edward Purcell, Ph.D. thesis (1938).

1946
"Radiological use of fast protons," R.R. Wilson, Radiology, 47:487–491 (1946).

1947
"I Absorption Spectrum of Hydrogen Peroxide, II Role of Hydro Peroxide in the Thermal Combination of Hydrogen and Oxygen," Roland B. Holt, Ph.D. thesis (1947).

1948
Neutron Scattering from Helium and Polarization of Neutrons and Protons by Scattering, Joseph V. Lepore, Ph.D. thesis (1948).

1950
"The Harvard 95 inch Cyclotron," W.G. Cross, L.L. Davenport, H.I Ewen, R.J. Grenzebach, R.B. Holt, L.S.Lavatelli, R.A. Mack, A.J.Pote, N.R Ramsey L.G Ritner, F. B. Robie and P.J. vanheerden, ONR report, NR-026-012 July 1950.

"Neutron Energy Distribution in Proton Bombardment of Beryllium," David Bodansky, Physical Review 80:481 (1950).

"Neutron Energy Distribution in Proton Bombardment of Beryllium," David Bodansky, Ph.D. thesis (1950).

"Proton-Proton Scattering at 100 MeV," R.W. Birge, 80:490 (1950).

"The Conservation of Energy and Momentum in Compton Scattering," William G. Cross, Ph.D. thesis (1950).

"Causes of abnormal efficiencies in Scintillation Counters," W.G. Cross, Phys. Rev. 78:185 9195.

1951
"Range and Straggling of Protons between 35 and 120 MeV," N. B1oembergen and P. J. Van Heerden. Phys. Rev. 83:561 (1951).

"Determination of Radioactivity by Solution in a Liquid Scintillator," M. S. Raben and N. B1oembergen. Science 114:363 (1951).

"Proportional Counter Study of Reported Proton-Electron Emission from Aluminium," Ann Birge, Ph.D. thesis (1951).

"Proton-Proton Scattering at 100 MeV," Robert Birge, Ph.D. thesis (1951).

Protonproton Scatering at 105 and 75 MeV, R.W. Birge, U.E. Kruse, and Norman F. Ramsey, Phys Rev 83:561 (1951).

"Neutron Energy Distributions in Proton Bombardment of Be and C at 100 MeV," David Bodansky and Norman F. Ramsey, Physical Review 82:831–836 (1951).

"Radiation from Galactic Hydrogen at 1420 Megacycles per Second," Harold Ewen, Ph.D. thesis (1951).

"Excitation Functions for Nuclear Reactions in the Range 0-100 MeV," Norton Hintz, Ph.D. thesis (1951).

"The Law of Photoelectric Absorption," Leo Lavatelli, Ph.D. thesis (1951).

"Stacked Foil tgechnique for Nuclear Reactions with Internal Cyclotron Beam," N.M. Hintz, Phys.Rev, 82:304A. (1951).

"Excitation Functions with an Internal Cyclotron Beam," N.M. Hintz, 83, 185 (1951).

"A Dee Biassing System for a Frequency Modulated Cyclotron," L.L Davenport, L. Lavatelli, R.A. Mack, A.J.Pote and N.F. Ramsey, Rev. Sci. Instr. 22, 601, (1951).

1952
"Excitation in. High Energy Nuclear Reactions, N. M.Hintz, Phys. Rev.86:1042 (1952).

"Excitation Functions to 100 MeV, N. M. Hintz and N. F. Ramsey, Phys. Rev. 88:19 (1952).

"Measurement of Short Beta-Decay Lifetimes," R. B. Holt, R. N. Thorn and R. W. Waniek, Phys.Rev. 87:378 (1952).

"High Energy Nuclear Reactions and the Goldberger Model," J. W. Meadows, Phys. Rev. 88:143 (1952).

"Scattering of Protons by the Loosely Bound neutron in Beryllium," K. Strauch and J. A. Hofmann, Phys. Rev. 86:563 (1952).

1953
"Nuclear Transmutations induced in Photographic Emulsions," Richard G. Seed, Ph.D. thesis (1953).

"Neutrons Ejected from Neutrons by 50 MeV Protons," Joseph A. Hofman, Ph.D. thesis (1953).

"On the Interaction of 95 MeV Protons with D, 1i, Be, C, AI, Cu and Pb," J. A. Hofmann and K. Strauch, Nuclei. Phys. Rev. 90:449 (1953).

"The Electric Scattering of Deuterons," U. E. Kruse, N. F. Ramsey and B. J. Malenka, Phys. Rev. 89:655. (1953).

"The Electric Scattering of the Polarizable Deuteron," B. J. Malenka, U. E. Kruse and N. F. Ramsey, Phys. Rev. 91:1165 (1953).

"Theory of High-Energy Deuteron Pickup," B. J. Malenka, Phys. Rev. 92:516 (1953).

"Excitation Functions for Proton-Induced Reactions with Copper," J. W. Meadows, Phys. Rev. 91:885 (1953).

"Polarizability of the Deuteron," N. F. Ramsey, B. J. Malenka and U. E. Kruse, Phys. Rev. 91:1162 (1953).

"Nuclear Configurations Inferred from High-Energy Pickup Deuteron Distributions," W. Selove, Phys. Rev. 92:1328 (1953).

"Neutron-Proton Scattering near 180° at 93 MeV," W. Selove, K. Strauch and F. Titus, Phys. Rev. 92:724 (1953).

"On the Detection of High Energy Particles with a Fast Coincidence System," K. Strauch, Rev. Sci. Instr. 24:283 (1953).

"A Method of Measuring Short Betz-Decay Lifetimes," R. N. Thorn and R. W. Waniek, Rev. Sci., Instr. 24:391 (1953).

"Nuclear Stars in Emulsions," R. Waniek and Taiichiro Ohtusuka, Phys. Rev. 89:882 (1953).

"Angular Distribution of Particles from Stars," R. W. Waniek and T. Ohtusuka, Phys. Rev. 89:1307 (1953).

"The Energy Spectrum of Particles from Stars," R. W. Waniek and T. Ohtsuka, Phys. Rev. 91:1574 (1953).

1954

"Total Cross Sections of Liquefied Gases for High Energy Neutrons," Peter Hillman, Ph.D. thesis (1954).

"Neutron-Proton Scattering at 91 MeV," Ralph Stahl, Ph.D. thesis (1954).

"Elastic Scattering of 95 MeV Protons by Deuterons," John Teem, Ph.D. thesis (1954).

V. Culler and R. W. Waniek. Total Cross Sections for High-Energy Neutrons. Phys. Rev. 95:585 (1954).

"Total Cross Sections of Liquefied Gases for High-Energy Neutrons," P. Hillman, R. H. Stahl and N. F. Ramsey, Phys. Rev. 96:115 (1954).

"Polarization in High Energy p-n-n Douple Scattering," P. Hillman, V. Culler and N. F. Ramsey, Phys. Rev. 95:462 (1954).

Proton-Proton Scattering from 40 to 95 MeV," U. E. Kruse, J. M. Teem and N. F. Ramsey, Phys. Rev. 94:1795 (1954).

"Polarization in High-Energy Elastic Nucleon-Nucleus Scattering," B. J. Malenka, Phys. Rev. 95:522 (1954).

"A DC Comparison Radiometer," W. Selove, Rev. Sci. Instr. 25:120 (1954).

"Interactions of High-Energy Neutrons in Molybdenum," E. G. Silver and R. W. Waniek, Phys. Rev. 95:586 (1954).

"On the Use of Nuclear Research Emulsions with Embedded Wires," E: G. Silver and R. W. Waniek, Rev. Sci. Iustr. 25:1119 (1954).

"Neutron-Proton Scattering at 91 MeV," R. Stahl and N. F. Ramsey, Phys. Rev. 96:1310 (1954).

"Scattering of 96 MeV Protons from Light Nuclei," K. Strauch and W. F. Titus, Phys. Rev. 95:854 (1954).

"An Efficient Cryostat for Producing Temperatures between 4° and 80° K: The Production of Liquid Hydrogen Targets Using Liquid Helium," C. A. Swenson and R. H. Stahl, Rev. Sci. Instr. 25:608 (1954).

1955

"The Scattering of 96 MeV Protons from Several Nuclei," Walter F. Titus, Ph.D. thesis (1955).

"Proton-Proton Scattering from 40 to 95 MeV," Ulrich E. Kruse, Ph.D. thesis (1955).

"Energy Spectrum of π mesons Produced by 2.2 Gev Protons on Beryllium," Frederic L. Nieman (1955).

"Yields of Mercury Isoptopes from Proton Bombardment of Gold," Daniel K. Butler, Ph.D. thesis (1955).

"Total Cross Sections for High-Energy Neutrons," V. Culler and R. W. Waniek, Phys. Rev. 99:740 (1955).

"Bilateral Development of Thick Nuclear Emulsions," R. Fox. and R. W. Waniek, Nucleonics 13:52 (1955).

"Magnetic Analysis of Scattered Particles," H. P. Furth, Rev. Sci. Instr 26:1097 (1955).

"Application of High Magnetic Fields to Nuclear Track Analysis and Solid State Research," H. P. Furth and R. W. Waniek, Nuovo Cimento 2, 1350 (1955).

"(p,a) and.(p,ab) Reactions at 100 MeV," J. W. Meadows, Phys. Rev. 98:744 (1955).

"Angular Distribution of Pickup Deuterons for 95 MeV Protons on Carbon, and Implications as to Internal Interactions in Carbon," W. Selove, Phys. Rev. 98:208 (1955).

"Double Scattering Experiment with 96 MeV Protons," K. Strauch, Phys. Rev. 99:150 (1955).

1956

"Fine structure in the nuclear photoeffect," Richard Wilson,Phys.Rev. 104: 1424 (1956).

"Interactions of K^- Mesons in Nuclear Emulsions," D. M. Fournet and M. Widgoff, Phys. Rev. 102:929 (1956).

"Production and Use of High Transient Magnetic Fields," H. P. Furth and R. W. Waniek, Rev. Sci. Instr. 27:195 (1956).

"High-Field Longitudinal Magnetoresistance of Germanium," H. P. Furth and R. W. Waniek, Fhys. Rev. 104:343 (1956).

" Interpretation of the Be^9(p,d) Reaction at energies of 5 to 30 MeV," S. Glashow and W. Selove, Phys. Rev. 102:200 (1956)

"Proton-Proton Scattering from 40 to 95 MeV," U. E. Kruse, J. M. Teem and N. F. Ramsey, Phys. Rev. 101:1079 (1956).

"Excitation Functions and Yield Ratios for the Iosmeric Pairs $Br^{80},^{80}m$, $Co^{58},^{58}m$, and $Sc^{44},^{44}m$ formed in (p,pn) Reactions," J. W. Meadows, R. M. Diamond and R. A. Sharp, Phys. Rev. 102:190 (1956).

"Characteristics of K^{+-} Particles," D. M. Ritson, A. Pevsner, S. C. Fung, M. Widgoff, G.T. Zorn, S. Goldhaber and G. Goldhaber, Phys. Rev. 101:1085 (1956).

"Interpretation of the H^3(p,n) Reaction," W. Selove, Phys. Rev.103:136 (1956).

"Table of Properties of the 'Elementary' Particles," A. M. Shapiro, Revs. Modern Phys. 28:164 (1956).

"Interactions of 1.4 BeV K^- Mesons with Helium Nuclei, A. M. Shapiro, Bull. Am. Phys. Soc. (II) 1:72 (1956).

"Direct Excitation of Nuclear Energy States in Carbon by 96 MeV," K. Strauch and F. W. Titus, Protons. Phys. Rev. 103:200 (1956).

"Inelastic Scattering of 96 MeV Protons," K. Strauch and F. W. Titus, Phys. Rev. 104:191 (1956).

"Scattering of K Particles," M. Widgoff, A. M. Shapiro, R. Schluter, D. M. Ritson, A. Pevsner and V. P. Henri, Phys. Rev.104:811 (1956).

"Fine Structure in the Nuclear Photoeffect," Richard Wilson. Phys. Rev. 104:1424 (1956).

"Mean Life Time of Negative K Mesons," M. Widgoff, Phys. Rev.102:1927 (1956).

1957

"Gamma-Ray Spectra from Elements Bombarded by 24 to 107 MeV Protons," Vaughn E. Culler, Ph.D. thesis (1957).

"Some Features of Regenerative Deflection and their Application to the Harvard Synchrocyclotron," G. Calame, P.F. Cooper, Jr., S. Engelsberg, G.L. Gerstein, A.M. Koehler, A. Kuckes, J.W. Meadows, K. Strauch and R. Wilson, Nucl. Instr. 1:169 (1957).

"Charge Density in the Deuteron," J. Bernstein, Phys. Rev. 108:349 (1957).

"Final-State Interactions in the Total Cross Section for Deuteron Photodisintegration," J. Bernstein, Phys. Rev. 106:791 (1957).

"Scattering of K⁺ Particles from Protons and Deuterons," J. Bernstein, Phys. Rev. 105:1853 (1957).

"Some Features of Regenerative Deflection and their Application to the Harvard Synchrocyclotron," G. Calame, P. F. Cooper, Jr., S. Engelsberg, G. L. Gerstein, A. M. Koehler, A. Kuckes, J. W. Meadows, Jr., K. Strauch and R. Wilson, Nuclear Instr. 1:169 (1957).

"Production and Use of High Transient Magnetic Fields," H. P. Furth, M. A. Levine and R. W. Waniek, Rev. Sci. Instr. 28:949 (1957).

"Elastic Scattering of 96 MeV Protons," G. Gerstein, J. Niederer and K. Strauch, Phys. Rev. 108:427 (1957).

"Energy Spectrum of Charged Pions from 2.2 BeV Protons on Be," F. L. Niemann, J. K. Bowker, W. M. Preston and J. C. Street, Phys. Rev. 108:1331 (1957).

"Asymmetry of Low Energy Positrons from Muon Decay," I. A. Pless, A. E. Brenner, R. W. Williams, R. Bizzani, R. H. Hildebrand, R. H. Milburn, A. M. Shapiro, K. Strauch, J. C. Street and L. A. Young, Phys. Rev. 108:159 (1957).

"Energy Dependence of the K⁻ Meson Interaction Cross Section," M. Widgoff, A. Pevsner, D.,F. Davis, D. N. Ritson, R. Schluter, and H. P. Henri, Phys. Rev. 107:1430 (1957).

1958

"The nuclear photoeffect," G.R. Bishop and R. Wilson, in *Handbook of Physics* XLI, Springer Verlag (1958).

"Cross Sections and Asymmetries of Pickup Deuterons at 145 MeV," Paul F. Cooper, Ph.D. thesis (1958).

"Angular and Energy Distributions of Low Energy Protons from Targets Bombarded by High Energy Protons," Raymond Fox, Ph.D. thesis (1958).

"The Elastic Scattering of 160 MeV Protons by Nuclei of Adjacent Mass," George L. Gerstein, Ph.D. thesis (1958).

"Proton-proton scattering at energies from 46 to 147 MeV," J.N. Palmieri, A.M. Cormack, N.F. Ramsey and Richard Wilson, Annals of Physics,5:299 (1958).

"Search for Neutron Resonances and P-N asymmetries at 145 MEv," Stuart G. Carpenter, Ph.D. thesis (1958).

"Angular Distribution of Fragments from Fission of U^{238} and Th^{232} by 45, 80, and 155-MeV Protons," J. W. Meadows, Phys. Rev. 110:1109 (1958).

"Proton-Proton Scattering at Energies from 46 to 147 MeV," J. N. Palmieri, A. M. Cormack, N. F. Ramsey and R. Wilson, Annals of Physics 5,229 (1958).

"Measurements of the Interaction of 95 MeV Protons with H4," W. Selove and J. M.Teem, Phys. Rev. 112:1658 (1958).

"Evidence Against the Reaction $K^+ \rightarrow \mu^+ \mu^\circ$," V. P. Henri, and A. M. Shapiro, Phys. Rev. 110:591 (1958).

"Gyromagnetic Ratios of Short-lived Nuclear State with Fast Neutral Beams," R. W. Waniek, Phys. Rev. 110:1479 (1958).

1959

"Search for Resonance Structure of Neutron Cross Sections at 100MeV," S.G. Carpenter and Richard Wilson, Phys. Rev. 114:510 (1959).

"p-n Asymmetries at 143 MeV," S.G. Carpenter and Richard Wilson, Phys. Rev. 113:650 (1959).

"Polarized Proton-Proton Scattering at Energies of 46 to 147 MeV," Joseph N. Palmieri, Ph.D. thesis (1959).

"The Elastic Scattering of 146 MeV Polarized Protons by Deuterons," Herman Postma, Ph.D. thesis (1959).

"Elastic Scattering and Polarization of Protons by Helium at 147 and 66 MeV," A.M. Cormack, J.N. Palmieri, N.F. Ramsey and Richard Wilson, Phys. Rev. 115:59 (1959).

"p-p Triple Scattering at 143 MeV," C.F. Hwang, T.R. Ophel, E.H. Thorndike, Richard Wilson and N.F. Ramsey, Phys. Rev. Letts. 2:514 (1959).

"Scintillation from Liquid Helium," E. H. Thorndike and W. J. Shlafer, Review of Scientific Instruments 30:838 (1959).

"Coupled Errors on Scattering Phase Shifts," Douglas Miller, Nuclear Physics 14:238 (1959).

"Nuclear parameters in the scattering of nucleons by carbon,"Richard Wilson, Phys. Rev. 114:260 (1959).

"A Search for a Neutral Spin 1 Meson in the π mass range," Robert Ely, MIT Ph.D. thesis.

"p-n Asymmetries at 143 MeV.," S. G. Carpenter and R. Wilson, Phys. Rev. 113:650 (1959).

"Search for Resonance Structure of Neutron Cross Section at 110 MeV," S. G. Carpenter and R. Wilson, Phys. Rev. 114:510 (1959).

"Elastic Scattering and Polarization of Protons by Helium at 146 and 66 MeV," A. M. Cormack, J. N. Palmieri, N. F. Ramsey and R. Wilson, Phys. Rev. 115:599 (1959).

"Search for $\pi^\circ \rightarrow 3\,\gamma$," R. P. Ely and D. H. Frisch, Phys. Rev. Letters, 3:565 (1959).

"Production of Σ^+ Hyperons by 990 MeV Positive Pions in Liquid Hydrogen," A. R. Erwin, Kopp, A. M. Shapiro, Phys.Rev. 115:669 (1959).

"p-p Triple Scattering at 143 MeV," C. F. Hwang, T. R. Ophel, E. H. Thorndike, R. Wilson and N. F. Ramsey, Phys. Rev. Letters 2:310 (1959).

"Coupled Errors on Scattering Phase Shifts," D. Miller, Nuclear Phys. 14:238 (1959).

"Scintillations from Liquid Helium," E. H. Thorndike and W. J. Shlaer, Rev. Sci. Instr. 30:838, (1959).

"Nuclear Parameters in the Scattering of Nucleons by Carbon," R. Wilson, Phys. Rev. 114:260 (1959).

"Sea Level Cosmic Ray Mass Spectrum in the Interval $30m_e$ – 2000me," G. G. Fazio and M. Widgoff. Phys. Rev. 116:1263 (1959).

1960

"Nuclear Radii from Neutron Scattering," Richard Wilson, Nucl.Phys. 6:318 (1960).

"Small-angle Proton Scattering at 3 BeV," W.M. Preston, Richard Wilson and J.C. Street, Phys. Rev. 118:579 (1960).

"Experimental Status of the Nucleon-nucleon Interaction," R.Wilson, report to the London conference on nuclear forces and the few nucleon problem. Published in Nuclear Forces and the Few-Nucleaon Problem, Pergamon Press, 1960; Vol. I, pp. 47–64.

"Depolarization and Time Reversal in p–p Scattering at 142MeV," C.F. Hwang, T.R. Ophel, E.H. Thorndike and Richard Wilson, Phys.Rev.119:352 (1960).

"Cross section and Asymmetry in the Deuteron Pickup Reaction C^{12} (p,d)C^{11} at 145 MeV," P.F. Cooper, Jr. and Richard Wilson, Nucl. Phys. 15:373 (1960).

"Measurement of the rotation parameter R in proton-proton scattering at 140 MeV," E.H. Thorndike, J. LeFrancois and Richard Wilson, Phys. Rev. 120:1819 (1960).

"The Scattering of Polarized Neutrons by Protons at 128 MeV," Russell K. Hobbie, Ph.D. thesis (1960).

"The Depolarization in Proton-Proton Scattering at 142 MeV," Chester F. Hwang, Ph.D. thesis (1960).

"The Quasi-Elastic PP and PN Scattering of 145 MeV Protons in Deuterium," Arthur F. Kuckes, Ph.D. thesis (1960).

"Proton-Proton Triple Scattering: Depolarization at 98 MeV, and Rotation at 140 MeV," Edward H. Thorndike, Ph.D. thesis (1960).

"Proton-Proton Depolarization at 98 MeV," E. H. Thorndike and T. R. Ophel, Phys. Rev. 119:362 (1960).

"Neutron-Proton Polarization and Differential Cross Section at 128 MeV," Russell K. Hobbie and Douglas Miller, Physical Review 120:2201 (1960).

"Efficient Neutron Detector with a Stable Energy Threshold," Douglas Miller and Russell K. Hobbie, Review of Scientific Instruments 31:621 (1960).

"High Energy Neutron Beam of 45% Polarization," Douglas Miller and Russell K. Hobbie, Physical Review 118:1391 (1960).

"Elastic (p,α) and (p,d) scattering at 147 and 66 MeV," A.M. Cormack, J.N. Palmieri, H. Postma, N.F. Ramsey and Richard Wilson, in *Nuclear Forces and the Few-Nucleon Problem*, Pergamon Press, 1960, pp. 259–268.

"Proton Radiation Damage in Silicon Solar Cells. Nuclear Electronic Effects Program," W. L. Brown and G. L.Pearson, Fourth Triannual Technical Note; Period 1 March 1960 – 30 June 1960 Contract AF 33(616)-6235.

"Cross Section and Asymmetry in the Deuteron Pickup Reaction C^{12}(p,d)C^{11} at 145 MeV," P. F. Cooper, Jr., and R. Wilson, Nuclear Phys. 15:373 (1960).

"Elastic (p,α) and (p,d) Scattering at 147 and 66 MeV," A. M. Cormack, J. N.Palmieri, H. Postma, N.F.Ramsey and R.Wilson. Reprinted from *Nuclear Forces and the Few Nucleon Problem*, Vol. 1, Pergamon Press, 1960, pp.259.

"Small Angle p-α and p-p Scattering at 150 MeV," A. H. Cromer. In: *Nuclear Forces and the Few Nucleon Problem*, Griffith and Power Ed. Vol. 1, 221, Pergamon Press, London (1960).

"Quasi-Elastic Proton-Proton Scattering at 158 MeV," B. Gottschalk and K. Strauch, Phys. Rev. 120:1005 (1960).

"Neutron-Proton Polarization and Differential Cross Sections at 128 MeV," R. K. Bobbie and D. Miller, Phys. Rev. 120:2201 (1960).

"Depolarization and Time Reversal in p-p Scattering at 142 MeV," C. F. Hwang, T. R. Ophel, E. H. Thorndike and R. Wilson, Phy. Rev. 119:352 (1960).

"Efficient Neutron Detector with a Stable Energy Threshold," D. Miller and R. K. Hobbie, Rev. Sci. Instr. 31:621 (1960).

"High Energy Neutron Beam of 45% Polarization," D. Miller and R. K. Hobbie, Phys. Rev. 118:1391 (1960).

"Small-Angle Proton Scattering at 3 BeV," W. M. Preston, R. Wilson and J. C. Street, Phys. Rev. 118:579 (1960).

"Measurement of the Rotation Parameter R in Proton-Proton Scattering at 140 MeV," E. H. Thorndike, J. Lefrancois, and R. Wilson, Phys. Rev. 120:1819 (1960).

"Proton-Proton Depolarization at 98 MeV," E. H. Thorndike and T. R. Ophel, Phys. Rev. 119:362 (1960).

"Experimental Status of the Nucleon-Nucleon Interaction," R. Wilson. Reprinted from *Nuclear Forces and the Few-Nucleon Problem*, Vol. 1, Pergamon Press, pp. 47–64 (1960).

"Nuclear Radii from Neutron Scattering," R. Wilson, Nuclear Phys.16:318 (1960).

1961

"Measurement of the R Parameter for Proton-Deuteron Elastic Scattering and Proton-Neutron Scattering in Deuterium," Jacques Le Francois, Ph.D. thesis (1961).

"Elastic Scattering of 146 MeV Polarized Protons by Deuterons, H. Postma and Richard Wilson, Phys. Rev. 121:1229 (1961).

"Asymmetry in 143 MeV pn Scattering," A.F. Kuckes and Richard Wilson, Phys. Rev. 121:1226 (1961).

"On the Deuteron as a Free Nucleon Target at 145 MeV," A.F.Kuckes, P.F. Cooper, Jr., and Richard Wilson, Annals of Physics 15:193 (1961).

"Diffusion Elastique des Nucleons par les Noyaux Legers (D, He, C) Comparison avec le Diffusion Nucleon-nucleon," Richard Wilson, J. Physique 22:610 (1961).

"The Measurement of Proton Inelastic Cross Sections between 77 MeV and 133 MeV," Raymond Goloskie, Ph.D. thesis (1961).

"Sensitive Monitor of Cyclotron Beam Position," B. Gottschalk, A.M. Koehler, D.J. Steinberg, The Review of Scientific Instruments, 32(6):744–745 (1961).

"The Use of the Bragg Peak of a Proton Beam for Intracerebral Lesions" [Abstract], R.N. Kjellberg, W.M. Preston, Excerpta Medica Int. Cong. Series (Second Int. Cong. of Neurological Surgery), 36, E103, (1961).

"Stereotaxic Instrument for Use with the Bragg Peak of a Proton Beam," R.N. Kjellberg, A.M. Koehler, W.M. Preston, W.H. Sweet, presented at First Int. Symposium on Stereoencephalotomy, Philadelphia, PA (1961).

"Radiation Damage Experiments on Pb Infrared Detectors," R. R. Billups and W. L. Gardner, MIT Lincoln Lab report 26G-0002, 23 March 1961.

"Sensitive Monitor of Cyclotron Beam Position," B. Gottschalk, A. M. Koehler and D. J. Steinberg, Rev. Sci. Instr. 32:744 (1961).

Asymmetry in 143 MeV pn Scattering," A. F. Kuckes and R. Wilson, Phys. Rev. 121:1226 (1961).

"On the Deuteron.as a Free Nucleon Target at 145 MeV," A. F. Kuckes, R. Wilson and P. F. Cooper, Jr., Ann. Phy. 15:193 (1961).

"The Use of the Bragg Peak of a Proton Beam for Intracerebral Lesions," R. N. Kjellberg and W. M. Preston, Excerpta Med. Int Cong. Series 36 (1961).

"π^+ -P Scattering at 990 MeV," Kopp, A. M., Shapiro, A. R. Erwin, Phys. Rev. 123:301 (1961).

"Duty Cycle Improvement on the Harvard Synchrocyclotron," J. Lefrancois, Rev. Sci. Instr. 32:986(1961).

"Elastic Scattering of 146 MeV Polarized Protons by Deuterons," H. Postma and R. Wilson, Phys.Rev. 121:1229 (1961).

"An In-Phantom Radiation Detector," M. S. Potsaid and G. Irie, N. E. J. Mad. 265:1135 (1961).

"Asymmetric Fission of Bismuth," T. T. Sugihara, J. Roesmer and J. W. Meadows, Jr., Phys. Rev. 121:1179 (1961).

"Chapter IV. Techniques of High Energy Physics," M. Widgoff, Interscience Publishers Inc., New York (1961).

1962

"n-p triple scattering parameters R and A," R.A. Hoffman, J.Lefrancois, E.H. Thorndike and Richard Wilson, Phys. Rev. 125:973 (1962).

"Quasi-Elastic Proton-Proton Scattering in Lig Nuclei," Bernard Gottschalk, Ph.D. thesis (1962).

"The Measurement of the Triple Scattering Parameter for Elastic and Quasi-Free Events of 137-1/2 MeV Protons on Deuterium," Richard Hoffman, Ph.D. thesis (1962).

"Tripple Scattering Parameter D_t for the Neutron-Proton Interaction," Potatlal M. Patel, Ph.D. thesis (1962).

"Slightly Inelastic Proton-Deuteron Scattering at 158 MeV," Douglas G. Stairs, Ph.D. thesis (1962).

"Measurement of the Triple Scattering Parameter D_t in the Free n-p System," P.M. Patel, A. Carroll, N. Strax, and D. Miller, Physical Review 8:491 (1962).

"Nucleon From Factors in the Helicity Representation," Douglas G. Miller, Physical Review 127:1365 (1962).

"The Bragg Peak of a Proton Beam in Intracranial Therapy of Tumors, R.N. Kjellberg, W.H. Sweet, W.M. Preston, A.M. Koehler, Trans. Am. Neurological Assoc., 87, 216-218 (1962).

"Effets d'une radiation localise de protons sur les potentiels electro-physiologiques evoques dans le ganglion geniculatum laterale du chat," J. Seigfried, F.R. Ervin, A.M. Koehler, R.N. Kjellberg, Helv. Physiol. Acta, 20, C83-C84 (1962).

"Stereotaxic Instrument for Use with the Bragg Peak of a Proton Beam," R.N. Kjellberg, A.M. Koehler, W.M. Preston, W.H. Sweet, Confin. Neurol., 22: 183–189 (1962).

"Biological and clinical studies using the Bragg peak of the proton beam," R.N. Kjellberg, A.M. Koehler, W.M. Preston, W.H. Sweet, presented at Second Int. Cong. of Radiation Research, Harrogate, England (1962).

"A Measurement of Neutron Proton Polarization at 126 MeV," Alan S. Carroll, Ph.D. thesis (1962)

"Corrections to the Impulse Approximation in Scattering from Deuterons," A. E. Everett, Phys. Rev. 126,. 831 (1962).

"The Relative Biologic Effectiveness of Extremely Energetic Protons and Alphas," R. Fix, J. McConomy, J. Wall and C. K. Levy, Tech. Doc. Rept. No. MDC–TDR–62–3 (1962).

"Low-Energy Proton Production by 160 MeV. Protons," R. Fox and N. F. Ramsey, Phys. Rev. 125:1609 (1962).

"Measurement of Proton Inelastic Cross Sections between 77 MeV and 133 MeV," R. Goloskie and K. Strauch, Nuclear Phys. 29:474 (1962).

"n-p triple Scattering Parameters R and A," R. A. Hoffman, J. Lefrancois, E. H.Thorndike and R. Wilson, Phys. Rev. 125:973 (1962).

"Stereotaxic Instrument for Use with the Bragg Peak of a Proton Beam," R. N. Kjellberg, A. M. Koehler, W. M. Preston and W. H. Sweet, Confin. Neurol. 22:183 (1962).

"Biological and Clinical Studies Using the Bragg Feak of the Proton Beam," R. N. Kjellberg, A. M. Koehler, W. M. Preston and W. H. Sweet, Sec. Int. Congo, Rad. Res. Harrogate, England. (1962).

"The Bragg Peak of a Proton Beam in Intracranial Therapy of Tumors. Trans," R. N. Kjellberg, W. H. Sweet, W. M. Preston and A. M. Koehler, Amer. Neurol. Ass. 87:216 (1962).

"Experimental Techniques for the Measurement of Nuclear Secondaries, from the Interactions of Protons of a Few Hundred MeV," F. C. Maienschein, Proc. Svmposium on the Protection Against Radiation Hazards in Space, Gatlinburg,Tenn. Nov. 1962; TIG-7652, Book 2.

"Measurement of the Triple Scattering Parameter D_t in the Free n-p System," P. M. Patel, A. Carroll, N. Strax and D. Miller, Phys. Rev. Letters 8, 491 (1962).

"Effets d'une Radiation Localisee de Frotons sur les Fotentiels Electro-physiologiques Evoques dans Ie Ganglion Geniculatum Laterale du chat," J. Seigfried, F. R. Ervin, A. M. Koehler and R. N. Kjellbeig, Helv. Physiol. Acta 20, C83 (1962).

"Extrapolation of Proton-Proton Scattering Data to the One-Pion Exchange Pole," E. H. Thorndike, Phys. Rev. 127,251 (1962).

1963

"Recoil Properties of Fission Products from Uranium bombarded with 150 MeV Protons," Victor E. Nioshkin. Clark University Ph.D. thesis.

"Measurement of the Triple-Scattering Parameter A in Proton-Proton and Proton-Carbon Scattering at 139 MeV," Stanley Hee and E. H. Thorndike, Physical Review 132:744 (1963).

"Slightly Inelastic Proton-deuteron Scattering," D.G. Stairs, Richard Wilson and P.F. Cooper, Jr., Phys. Rev. 129:1672 (1963).

"Measurement of the Triple Scattering Parameter R' in Proton-proton Scattering at 137 1/2 MeV," Stanley Hee and Richard Wilson, Phys. Rev. 132:2236 (1963).

"Proton-deuteron Elastic Triple Scattering at 140 MeV," R.A.Hoffman, J.

Lefrancois, E.H. Thorndike and Richard Wilson, Phys. Rev. 131:1671 (1963).

"Quasi Free Proton-neutron and Proton-proton Scattering at 140MeV," J. Lefrancois, R.A. Hoffman, E.H. Thorndike and Richard Wilson, Phys. Rev. 131, 1660 (1963). 127: 87-97.

"Measurement of the Triple Scattering Parameter A in Proton-proton Scattering at 139 MeV," Stanley Hee and E.H Thorndike, Phys. Rev. 132:744–749 (1963).

"A Measurement of Neutron-Proton Polarization at 126 MeV," Alan Carroll, Ph.D. thesis (1963).

"Theory of Inelastic Proton-Deuteron Scattering," A. H. Cromer, Phys. Rev. 129:1680 (1963).

"Proton-Proton Bremsstrahlung of the p-p Interaction," A. H. Cromer and M. I. Sobel, Phys. Rev. 132:2698 (1963).

"Interpretation of Quasi-Free Proton-Deuteron Scattering," A. H. Cromer and E. H. Thorndike, Phys. Rev. 131:1680 (1963).

"Measurement of the Triple Scattering Parameter A in Proton-Proton and Proton-Carbon Scattering at 139 MeV," S. Hee and E. H. Thorndike, Phys. Rev. 132:744 (1963).

"Measurement of the Triple Scattering Parameter R' in the Proton-Proton Scattering at 137½ MeV," S.Hee and R. Wilson. Phys. Rev. 132:2236 (1963).

"Proton-Deuteron Elastic Triple Scattering at 140 MeV," R. A. Hoffman, J. Lefrancois and E. H. Thorndike, Phys. Rev. 132:1671 (1963).

"Small Lesions in the Thalamus in Parkinsonian Patients," R. N. Kjellberg, Surgical Forum XIV, 435 (1963).

"Quasi-Free Proton-Neutron and Proton-Proton Scattering at 140 MeV," J. Lefrancois, R. A. Hoffman, E. H. Thorndike and R. Wilson, Phys. Rev. 131:1660 (1963).

"Effects of X-rays and High Energy Protons on the Sensitive Plant (mimosapidica)," C. K. Levy and K. Kerl, Fed. Proc. 22:590 (1963).

"The Depth-Dose Distribution Produced in a Spherical Water Filled Phantom by the Interactions of a 160 MeV Proton Beam," F.C. Maienschein and T. V. Blosser, ORNL-3457 (1963).

"Neutron Detection and Analysis Using a Scintillation Bubble Chamber," R. C. Minehart and R. H.Milburn, Nuclear Instr. and Methods 25:13 (1963).

"Normalization of p-p, p-d and p-He Differential Cross Sections Measured at the Harvard Cyclotron Laboratory," J. N. Palmieri, R. Goloskie and A. M. Cormack, Phys. Letters 6:289 (1963).

"Solid 'Phantom' Dosimeters," M. S. Potsaid. A Chapter in "The Encyclopedia of X-rays and Gamma Rays," Ed. George Clark Theingold, New York, 279 (1963).

"Slightly Inelastic Proton-Deuteron Scattering," D. G. Stairs, R. Wilson and P. F. Cooper, Jr., Phys. Rev. 129:1672 (1963).

"The Nucleon-Nucleon Interaction: Experimental Aspects," R. Wilson, Wiley, Interscience (1963).

1964

"Triple Scattering Near 140 MeV: A in Proton-Proton and Proton-Carbon Scattering and R' in Proton-Proton Scattering," Stanley Y. Hee, Ph.D. thesis (1964).

"Proton-Proton Bremsstrahlung," Michael Sobel, Ph.D. thesis (1964).

"Measurement of the Triple Scattering Parameter D_t in the Free n-p System at 170°," W. G. Collins, JR. and D. G. Miller, Physical Review 134:B575 (1964).

"Enzyme Inactivation by High Energy Protons," D.H. Thomson, K. Stratton, R. Tym, A.M. Koehler, Abstract in Radiation Research, 22:243 (1964).

"Free Radical Formation in Proton Irradiated Enzymes" [Abstract], K. Stratton, A.M. Koehler, Abstract in Radiation Research, 22:240 (1964).

"Intracranial lesions Made by the Bragg Peak of a Proton Beam," R.N. Kjellberg, A.M. Koehler, W.M. Preston, W.H. Sweet, in *Response of the Nervous System to Ionizing Radiation*, ed, T.J. Haley, R.S. Snider, Little, Brown & Co., Boston, 36–53 (1964).

"Facilities for Biological and Medical Uses of the Harvard Cyclotron," A.M. Koehler, Presented at Conference on High Energy Cyclotron Improvement, Williamsburg, Va (1964).

"Problèmes a Petit Nombre de Nucleons," Richard Wilson, in Comptes Rendus du Congress International de Physique Nucleaire, P. Gugenburger (ed.), Paris, 1964.

"The Effect of Angular Momentum on Isomeric Cross section Ratios in Compound Nuclear Reactions," Norman A. Dudey, Clark University Ph.D. thesis.

"A Radiochemical Study of Bismuth Fission Products induced with 38 and 58 MeV Protons," Josef Roesmer, Clark Univerity Ph.D. thesis "Nucleon-Nucleon Scattering", Richard Wilson., Wiley-Interscience.

T. V. Blosser, F. C. Maienschein and R. M. Freestone, Jr. Health Phys. 10:743 (1964).

"Small Angle Polarization in n-p Double Scattering," A. S. Carroll, D. G. Miller, P. M. Patel and N. Strax, Phys. Rev. 134:595 (1964).

"Measurement of the Triple Scattering Parameter D_t in the Free n-p System at 170°," W. G. Collins, Jr. and D. G. Miller, Phys. Rev. 134:575 (1964).

"Studies of High Energy Electron (10–60 MeV) and Proton (20–130 MeV) Radiation Damage in Silicon and Germanium," J. C. Corelli, L.J. Cheng, C. A. Taylor and O. H. Morrill, Effets des: Rayonnements sur les Semiconducteurs, Paris (1964).

"Analysis of Nucleon-Nucleus Scattering," A. H. Cromer and J. N. Palmieri, Ann. Phys. (N.Y.). 30, 32 (1964).

"Carrier Lifetime Studies in Electron (10–55 MeV) and Proton (25–110 MeV) Irradiated Germanium," J. E. Fischer and J. C. Corelli, Jour. of Applied Phys. 35:3531 (1964).

"Total" Proton-Proton Cross Sections," R. Goloskie and J. N. Palmieri, Nuclear Phys. 55:463 (1964).

"Quasi-Elastic p-p Scattering in Fluorine and Some 'Medium' Nuclei," B. Gottschalk, K. Strauch and K. H.Wang, Reprinted from the Proceedings of the International Conference on Nuclear Physics, Vol. 2,Communications; Centre National de la Recherche Scientifique, Paris, 324 (1964).

"Performance Characteristics of Modular Nanosecond Circuitry Employing Tunnel Diodes," N. W. Hill, ORNL-3687, October (1964).

"Intracranial Lesions Made by the Bragg Peak of a Proton Beam," R. N., Kjellberg, A. M. Koehler, W. M. Preston and W. H. Sweet. In Haley and Snider's *Response of the Nervous System to Ionizing Radiation.* (Little, Brown, Boston) 36 (1964).

"Calibration of 30 cm Faraday Cup," J. N. Palmieri and R. Goloskie, Rev. Sci. Instr. 35:1023 (1964).

"Total Nucleon-Helium Cross Section near 147 MeV," J. N. Palmieri and R. Goloskie, Nuclear Phys. 59:253 (1964).

"Differential Cross Sections by Flight-Time Spectroscopy for Proton Production in Reactions of 160 MeV Protons on Nuclei," R. W. Peelle. Second Svmposium on Protection Against Radiation Hazards in Space, Gatlinburg, Tenn. Oct, 1964 NASA SP-71.

"Energy Dependence of Proton Irradiation Damage in Silicon," W. Rosenzweig, F. M. Smits and W. L. Brown, J. Appl. Phys. 35(9), Sept. (1964).

"Measurement of the Intensity of the Proton Beam of the Harvard University Synchrocyclotron for Energy Spectra Measurements of Nuclear Secondaries," R. T. Santoro and R. W. Peelle, NRNL-3505 (1964).

"A Multicrystal Gamma-Ray Spectrometer with Time-of-flight Rejection of Neutron-Induced Background," R. J. Scroggs, IEEE Trans. Nucl. Sci. NS-11:365 (1964).

"Small-Angle Scattering of 143 MeV Polarized Protons," D. J. Steinberg, J. N. Palmieri and A. M. Cormack, Nuclear Phys. 56:46 (1964).

"Free Radical Formation in Proton Irradiated Enzymes," K. Stratton and A. M. Koehler, Radiation Research 22:240 (1964).

"Enzyme Inactivation by High Energy Protons," D. H. Thomson, K. Stratton, R. Tym and A. M. Koehler, Radiation Research 22:243 (1964).

"Spectra of Gamma Rays Produced by Interaction of 160 MeV Protons with Be, C, O, Al, Co, and Bi," W. Zobel, Proc. Second Symp Protection Against Radiation in Space, NASA SP-71, 341 Oct. (1964).

"The Influence of High Energy Particle Radiation in Space on Different Kinds of Transistors," William C. Honnaker and Floyd R. Bryant, Atompraxis, May (1964).

"Space Particle Environment and its Effects on Microelectronics," E. Rind and F. R. Bryant, Electronics Systems Symposium and Exhibit on Electrotechnology, April (1964).

"Experimental Investigation of Simulated Space Radiation Effects on Micro-electronics," E. Rind and F. R. Bryant, IEEE International Conf., March (1964).

"The Effects of Dose, Dose Rate and Quality of Radiation on the Dynamics and Survival of the Spermatogenic Population of the Mouse," E. F. Oakberg, Suppl. to Jap. J. Genet. 40:119 (1964).

1965

"Bragg Peak Pituitary Destruction for Diabetic Retinopathy," R.N. Kjellberg, R.A. Field, J.W. McMeel, W.H. Sweet, Excerpta Medica Int. Cong. Series, pp. 93–102 (1965).

"Hemorrhagic Diabetic Retinopathy: Treatment by Pituitary Stalk Section or by Pituitary Irradiation with Protons at the Bragg Peak," R.N. Kjellberg, A.M. Koehler, W.M. Preston, W.N. Sweet, Neurologia Medico-Chirurgica, 7:1–13 (1965).

"Measurement of Cross Sections with Neutrons as Targets," M.W. Shapiro, A.M. Cormack, A.M. Koehler, The Physical Review, 138 (4B), B823-B830 (May 1965).

"Radiation Damage of Mylar and H-film," A.M. Koehler, D.F. Measday, D.H. Morrill, Nuc. Inst. and Meth., 33:341–342 (1965).

"Range of Protons in Human Skullbone," A.M. Koehler, J. Dickinson, W.M. Preston, Radiation Research, 26(3):334–342 (1965).

"The Energy and Momentum Dependence of NP Charge Exchange Scattering," Richard Wilson, Annals of Physics 32, 193 (1965).

"A study of Symmetric Fission of Light Elements," Hiroshi Baba, Clark University Ph.D. thesis.

"Isomeric Cross Section Ratios in the (p,pn) and Heavy Ion Neutron Transfer Reactions," Christopher Gatrousis, Clark University Ph.D. thesis.

"Elastic and Inelastic Scattering of High Energy Protons from Nucleii," Philip G. Roos, MIT Ph.D. thesis.

"A Measurement 01 Proton-Proton Bremsstrahlung at 150 MeV," B. Gottschalk, W.J. Sh1aer and K. H. Wang, Pys. Rev. 16, 294 (1965).

"Bragg Peak Proton Pituitary Destruction for Diabetic Retinopathy," R. N. Kjellberg, R. A. Field, J. W. McMeel and W. H. Sweet, Third Int. Cong. Neurol. Surg., Copenhagen, August (1965).

"Hemorrhagic Diabetic Retinopathy: Treatment by Pituitary Stalk Section or by Pituitary Irradiation with Protons at the Bragg Peak," R. N. Kjellberg, A. M. Koehler, W. M. Preston and W. H. Sweet, Neurologic Medico-Chirurgica (Eng. edition) 7:1 (1965).

"Radiation Damage of Mylar and H-Film," A. M. Koehler, D. F. Measday and D. H. Morrill, Nuclear Instr. and Methods 33, 341 (1965).

"Dose-rate Phenomena and Relative Biological Effectiveness of High Energy Protons in Ascites Tumor Cells," T. R. Marcus and C. K. Levy, Rad. Research 25(1) May (1965).

"Loss of Protons by Nuclear Interactions in Sodium Iodide Crystals," D. F. Measday, Nuclear Instr. and Methods, 34,353 (1965).

"Processes for Obtaining Conducting Areas on Thin Film: Application Ionization Chambers," D. F. Measday, Rev. Sci. Instr. 36:1263 (1965).

"Proton Dosimetry Using the HAP System," M. S. Potsaid, T. Umezawa, S. Shibata and A. M. Koehler. Report to NASA on Research Grant NsG 719: Solid Chemical Radiation Dosimeter, p. 1, December 20, 1965.

"Polymer Solutions as Radiation Dosimeters," R. M. Raju, Phys. Med. Biol. 10:515 (1965).

"The Space, Time and Energy Distributions of the Proton Beam of the Harvard University Synchrocyclotron, R. T. Santoro, ORNL-3722 (1965).

"Treatment of Intractable Temporal Lobe Epilepsy by Sterotactic Amygdala Lesions," R. S. Schwab, W. H. Sweet, V. H. Mark, R. N. Kjellberg and F. R. Ervin, Trans. Amer. Neurol. Ass. 90:12 (1965).

"Limits on Parity and Time-Reversal Noninvariance in p-p Scattering," E. H. Thorndike, Phys. Rev. 138:586 (1965).

"Elastic Scattering of Protons from Be^9, Ca^{40}, Ni^{58}, Sn^{120} and Pb^{208}," N. S. Wall, Phys. Rev. 140:1237 (1965).

"Spectra of Gamma Rays Produced by the Interaction of 160 MeV Protons with Be, C. O, Al, Co and Bi," W. Zobel, ORNL-3506 April (1965).

"A Dose-Equated Manikin for Space Radiation Research," Janni, M. Schneider, B. Clark and P. Berger, AFWL-TR-65-95, August (1965).

"Isomeric Cross-Section Ratios in the (p, pn) and Heavy-Ion Neutron-Transfer Reaction," C. Gatrousis, US AEC Report NYO-1930-44 (1965).

"A Study of Symmetric Fission of Light Elements.," H. Baba, US AEC Report NYO-1930-43 (1965). Chem. Abst. 62, 12573F (1965).

"Cross-Section Ratios of Isomeric Nuclides Produced in Medium-Energy (α,xn) Reactions," J. T. Matsuo, J. M. Matuszek, N. D. Dudey and T. T. Sugihara, Phys. Rev. 139, B886 (1965).

"Statistical-Model Analysis of Isomeric Ratios in (α,xn) Compound Nuclear Reactions," N. D. Dudey and T. T. Sugihara, Phys. Rev. 139 B896 (1965).

"Recoil Properties of Fragments in 150 MeV Proton Fission of Uranium-238.

I Ranges and Kinetic Energies," V. E. Noshkin, Jr. and T. T. Sugihara, J. Inorg. Nucl. Chem. 27:943 (1965).

"Recoil Properties of Fragments in 150 MeV Proton Fission of Uranium 238. II Momentum Transfer and Excitation Energy," V. E. Noshkin, Jr. and T. T. Sugihara, J. Inorg. Nucl. Chem. 27:959 (1965).

"Measurements of Cross Sections with Neutron as Targets," M. Widgoff, Phys. Rev. 138, B823 (1965).

"Gatrousis. Isomeric Cross-Section Ratios in the (p,pn) and Heavy-Ion Neutron-Transfer Reaction," Christopher, US AEC Report NYO-1930–44 (1965).

"A Study of Symmetric Fission of Light Elements," Hiroshi Baba, Chem. Abst. 62, 12573f (1965).

"Recoil Properties of Fragments in 150 MeV Proton Fission of Uranium 238. I Ranges and Kinetic Energies," V. E. Noshkin, Jr. and T. T. Sugihara, J. Inorg. Nucl. Chem. 27:943 (1965).

"Recoil Properties of Fragment in 150 MeV Proton Fission of Uranium 238. II Momentum Transfer and Excitation Energy," V. E. Noshkin, Jr. and T. T. Sugihara, J. Inorg. Nucl. Chem. 27:959 (1965).

"Cross Section Ratios of Isomeric Nuclides in Medium-Energy (α,xn) Reactions," T. Matsuo, J. M. Matuszik, Jr. N. D. Dudey and T. T. Sugihara, Phys. Rev. 139:886 (1965).

"Statistical-Model Analysis of Isomeric Ratios in (α,xn) Compound Nuclear Reactions," N. D. Dudey and T. T. Sugihara, Phys. Rev. 139:896 (1965).

1966

"Enhancement of Whole-Body Lethality to X-ray and proton Irradiation by Pre Exposure of the Skin to Soft X-rays," Leo Fox, BU Ph.D. thesis.

"Effects of Dose and Dose rate of High energy Protons on Chromosome Breakage in Mammalian Cells," Theodore Marcus, BU Ph.D. thesis.

"Proton Proton Bremsstrahlung at 158 MeV," William J. Shlaer, Ph.D. thesis (1966).

"A Measurement of the Triple-scattering Parameter A_t for Free Neutron-proton Scattering," Norman Strax, Ph.D. thesis (1966).

"Effects of Radiation Mediating Agents on the Response of a Murine Ependymoma to Proton Radiation," K. Stratton, A. Anderson, A.M. Koehler, Radiology, 87(1):68–73 (1966).

"Loss of Charged Particles by Nuclear Interactions in Scinitllators," D. Measday, R. Schneider, Nucl. Inst. & Meth., 42:26–28 (1966).

"Proton-Proton Bremsstrahlung at 158 MeV," B. Gottschalk, W. J. Shlaer and K. H. Wang, Nuclear Phys. 75:549 (1966).

"The Enhancement of Lethality in High Energy Proton and X-irradiated Mice by Previous Exposure of Skin to Erythema Doses of Very Soft X-rays," C. K. Levy and Leo Fox, Abstract in Inter. Jour. Radiation Research (1966).

"Loss of Charged Particles by Nuclear Interactions in Scintillators," D. F. Measday and R. J. Schneider, Nuclear Instr. Methods 42:26 (1966).

"Monokinetic Neutron Beam in. the Range of 50 to 150 MeV," D. F. Measday, Nuclear Instr. Methods 40, 213 (1966).

"Neutron-Proton Differential Cross Section at 129 and 150 MeV," D. F. Measday, Phys. Rev. 142:584 (1966).

"Neutron Total Cross Sections for Neutrons, Protons, and Deuterons in the Energy Range of 90 to 150 MeV," D. F. Measd,ay and J. N. Palmieri, Nuclear Phys. 85:142 (1966).

"Neutron Total Cross Sections in the Energy Range 80 to 150 MeV," D. F. Measday and J. N. Palmieri, Nuclear Phys. 85:129 (1966).

"A Study of the (n,d) Reaction at 152 MeV," D. F. Measday, Phys. Letters 21:62 (1966).

"The C^{12} (p,pn) C^{11} Reaction from 50 to 160 MeV," D. F. Measday, Nuclear Phys. 78:476 (1966).

"The H^2(n,p)2n Reaction at 152 MeV," D. F. Measday, Phys. Letters 21:66 (1966).

"Space Chamber Neutron Spectrometer. Final Report," Marion C. Rinehart, Contract AT(30-1)3579. USAEC, Health and Safety Laboratory, New York, Philco-Ford SRS-TR 3148 (8 Dec. 1966).

"Proton Irradiation of an Experimental Murine Brain Tumor," K. Stratton and A. Anderson, Radiation Research, 27:538 (1966).

"Response or Murine Glioma Iso-transplants to Proton and X-radiation," K. Stratton and A. Anderson, Abst. 3rd. Int. Cong. of Radiation Research, Cortina, Italy, In vivo assay of Cell Survival and Sensitization by 5-iodo-2'deoxyuridine. (1966).

"Effects of Radiation Mediating Agents on the Response of a Murine Ependymona to Proton Irradiation," K. Stratton, A. Anderson and A. M. Koehler, Radiology, 87:68 (1966).

"Quasi-Free Scattering of 160 MeV Protons from Nuclei," N. S. Wall, Phys. Rev. 150:811 (1966).

"Differential Cross Sections for the Production of Protons in the Reactions of 160 MeV Protons on Complex Nuclei," R. W. Peelle, ORNL-3887 (1966). Submitted to Phys. Rev.

"Damage to Field Effect Transistors Under 22 and 128 MeV Proton Bombardment," Floyd R. Bryant and Carl L. Fales, TN-D-3630, (Oct. 1966).

"Some Effects of 138 MeV Protons on Primates," G. V. Dalrymple, I. R. Lindsay, J. J. Ghidoni, J. D. Hall, J. C. Mitchell, H. L. Kundel and I. L. Morgan, Rad. Res. 28:471 (1966).

"The Relative Biological Effectiveness of 138 MeV Protons as Compared to Cobalt 60 Gamma Radiation," G. V. Dalrymple, I. R. Lindsay, J. D. Hall, J. C. Mitchell, J.J. Ghidoni, H. L. Kundel and I. L. Morgan, Rad. Res. 28, 489 (1966).

"Proton Depth-Dose Dosimetry," J. C. Mitchell, G. V. Dalrymple, G. H. Williams, J. D. Hall and I. L. Morgan, Rad. Res. 28:390 (1966).

"The Effect of High Energy Proton Irradiation on the Cardiovascular System of the Rhesus Monkey," H. L. Kundel, Rad. Res. 28:529 (1966).

"Directional Neutron Detector for Space Research Use," E. L. Chupp and D. J. Forrest, IEEE Trans. on Nuclear Science NS-13, 1:468 (1966).

"Effects of High Energy Protons on Brain and Glioma Mice," S. H. M. Nystrom, Acta Radio. Stockholm, 5:133 (1966).

"Some Aspects of the Use of Protons in the Treatment of Experimental Brain Tumors," S. H. M. Nystrom, Naturwissenschaften, 53:159 (1966).

"Immediate Effect s of Protons on the Mammalian Brain," S. H. M. Nystrom, Naturwissenschaften 53:159 (1996).

"Experimental D8: Radiation in Spacecraft," M. Schneider and J. Janni, SSD-TR-115 (May 1966).

1967

"Immediate Physiological responses of Mamalian Sensory Motor systems," J. Leith, BU Ph.D. thesis.

"High Energy Proton Depth Dose Patterns," R. L. Tanner, N. A. Baily and J. W. Hilbert, Rad. Res. 32:861 (1967).

"Recent Results in Macro and Micro Dosimetry of High Energy Particulate Radiation," N. A. Baily, Proc. 1st Int. Symp. on the Biol. Interpretation of Dose from Accelerator Produced Rad., 329 TID.-4500, Conf.-670305 (1967).

"Effect of Dose Rate on Acute Mortality in Mice Exposed to 138 MeV Proton Irradiation," J. E. Traynor and E. T. Still, Presented at Radiation Research Society Meeting, May, 1967, San Juan, P. R.

"Dosimetry of Proton Beams using Small Silicon Diodes," A.M. Koehler, Radiation Research, supp. 7:53–63 (1967).

"Time-intensity Data in Solar Cosmic-ray Events: Biological Data Relevant to their Effects in Man," W.H. Sweet, R.N. Kjellberg, R.A. Field, A.M. Koehler, W.M. Preston, Radiation Research, supp. 7:369–383 (1967).

"Measurements of Spacecraft Cabin Radiation Distributions for the Fourth and Sixth Gemini Flights," J. Janni, AFWL-TR-65-149 (March 1967).

"A Correlation of Dosimetric Measurements with Charged Particle Environment of the Inner Van Allen Belt," A. Thede and G. Radke, (June 1967).

"The N-N interaction, 1967—An overview and conference summary," A.E.S.

Green, M.H. MacGregor and Richard Wilson, Rev. Mod. Phys. 39:498 (1967).

"High Energy Nucleon Scattering," Richard Wilson, Comments on Nuclear and Particle Physics I, 160 (1967).

"High Energy Proton Testing of Mariner IV Components," B. E. Anspaugh, Jet Propulsion Lab. Tech Memo 33–314, 1 (Jan. 1967).

"Potential Model Calculation of Proton-Proton Bremsstrahlung," A. H. Cromer, Phys. Rev. 158:1157 (1967).

"Bait-Shyness: A Test for Toxicity with N=2," J. Garcia, F. R. Ervin and R. Koellinger, Physchn. Sci. 7:245 (1967).

"Toxicity of Serum from Irradiated Donors," J. Garcia, F. R. Ervin and R. Koellinger, Nature 213:682 (1967).

"Proton Recoil Spectrometer for Neutron Spectra Between 50 and 450 MeV," W. A. Gibson, W. R. Burrus, J. W. Wachter and C. F. Johnson, Nucl., Instr and. Methods 45:29 (1967).

"Detection of 160 MeV Protons by a Lithium-Drifted Germanium Counter in a Side Entry Orientation," C. G. Gruhn, T. Kuo, B. Gottschalk, S. Kannenberg and N. S. Wall, Phys. Letters 24B, 266 (1967).

"Measurement of the Neutron-Proton and Neutron-Carbon Cross Sections at Electron Volt Energies," T. L. Houk and R. Wilson, Rev. Mod. Phys. 39: 546 (1967).

"Immediate Transient Responses to Ionizing Radiation," C. K. Levy, In: Current Topics of Radiation Biology, edited by Ebert and A. Howard. chapter 3, Vol. III, North Nolland Publishing Co., 44 pages (1967).

"Unbound Energy Levels in H^4," D. F. Measday and J. N. Palmieri, Phys Letters 25B, 106 (1967).

"The (n,p) and (n,d) Reactions at 152 MeV," D. F. Measday and J. N. Palmieri, Phys. Rev. 161:1071 (1967).

"Electron Spin Resonance Studies on Proton Irradiated, Ribonuclease and Lysozym," K. Stratton, Proceedings of the Workshop Conference on Space Radiation Biology, Berkelye, 1965. Radiation Research Suppl. 7:102 (1967).

"The Effect of Temperature on the Yield of Radiation-Induced Free Radicals in Several Dry Enzymes," K. Stratton, Radiation Research.31:585 (1967).

"Time Intensity Data in Solar Cosmic-ray Events: Biological Data Relevant to their Effects in Man," W. H. Sweet, R. N. Kjellberg, R. A. Field, A. M. Koehler and W. M. Preston, Radiation Research Supple, 7:369 (1967).

"Neutron and Proton Spectra from Targets Bombarded by 160 MeV Protons," J. W. Wachter, W. R. Burrus and W. A. Gibson, Phys. Rev. 161:971 (1967).

"Quasifree Scattering of 160-MeV Protons from Nuclei Phys," N.S. Wall and P. R. Roos, Rev. 159:1062 (1967).

"Scattering in Carbon," K. H. Wang, Nuclear Phys A90:83 (1967).

"Non-coplanarity in Proton-Proton Bremsstrahlung at 157 MeV," K. H. Wang, Nucl. Phys. A94:491 (1967).

"Cross Section for Formation of Be^7 by 20-155 MeV Proton-Induced Reactions in Carbon," I. R. Williams and C. B. Fulmer, Phys. Rev. 154:1005 (1967).

"n-n Scattering 1967. An Overview and Conference Summary," R. Wilson, Rev. Mod. Phys. 39:498 (1967).

"Gamma Rays from Bombardment of Li^7, Be, Be^{11}, C, O, Mg, Al, Co, Fe and Bi by 16- to 160 MeV Protons and 59 MeV Alpha Particles," W. Zobel, ORNL 4183 (November 1967).

"Relative Effectiveness of Gamma-Rays, X-rays, Protons, and Neutrons for Spermatogonical Killing," E. F. Oakberg, 1st. Inter. Symp. on the Biological Interpretation of Dose from Accelerator Produced Radiation. R. Wallace, ed., CONF-670305, pp. 174 (1967).

1968

"Total Cross Sections of Heavy Elements for 100 to 150 MeV Protons," Robert J. Schneider, Tufts University Ph.D. thesis.

"High energy Neutrojns and Solar Surface Nuclear Reactions," D.J. Forrest, UNH Ph.D. thesis.

"Quasi-Elastic Cluster Knock-Out Reactions using 100 MeV Protons," Susan Kannenberg, Northeastern University Ph.D. thesis.

"Low-energy Nucleon-nucleon Scattering, Charge Symmetry and Charge Independence," Comments on Nuclear and Particle Physics VII, 141 (1968).

"Effect of Proton Beam on Pituitary Function in Active Acromegaly," S. Field, B. Kliman, A.G. Frantz, R.N. Kjellberg, Abstract in Clinical Research, 16: 266 (1968).

"Neutron Total Cross-section in the Energy Range 100 to 150 MeV," R. Schneider, A. Cormack, Nuclear Physics A, 119:197–208 (1968).

"Proton Beam Irradiation of the Pituitary," R.N. Kjellberg, in *Clinical Endocrinology*, vol. 2., ch. 9, ed. E.B. Astwood, C.E. Cassidy, Grune and Stratton, N.Y., 103–109 (1968).

"Proton Beam Therapy in Acromegaly," R.N. Kjellberg, A. Shintani, A.G. Frantz, B. Kliman, New England J. of Medicine, 278:689–95 (Mar. 1968).

"Proton Radiography," A.M. Koehler, Science, 160:303 (1968).

Measurements of the Neutron-proton and Neutron-carbon Total Cross Sections at Electron-volt Energies," T.L. Houk and Richard Wilson, Rev.Mod. Phys. 39:546 (1967). Erratum: 40, 672 (1968).

"Satellite Instrumentation for Charged Particle Measurements: II Magnetic Analyzer for 0.1 to 1.0 MeV Electrons," G. Theodoridis, IEEE Trasn. on Nuclear Sci. NS-15(1) (1968).

"'Satellite Instrumentation for Charged' Particle Measurement: III Scintillation Spectrometer for Relativistic Electrons," G. Theodoridis, IEEE Trans. on Nuclear Sci, NS-15(1) (1968).

"Satellite Instrumentation for Charged Particle Measurements: IV Solid-State Spectrometers for 0.1 to 1.0 and 0.8 to 6.0 MeV Protons," G. Theodoridis, IEEE Trans. on Nuclear Sci. NS-15(1) (1968).

"Upper Limit to the Solar Neutron Flux in Energy Interval 20–200 MeV," D. J. Forrest artd E. L. Chupp, Special Meeting of the A.A.S., Solar Astronomy, Feb. 1–3, 1968, Tucson, Ariz.

1969

"Clinical Studies on Stereotactic Proton Hypophysectomy of Kjellberg's Method – Its Indication on Acromegaly," A. Shintani, R.N. Kjellberg, Abstract in Brain and Nerve, 21:57–67 (1969).

"A Comparison of Methods of Treatment for Diabetic Retinopathy," A.M. Koehler, R.N. Kjellberg, J.W. McMeel, in *Symposium on the Treatment of Diabetic Retinopathy*, ed. M.F. Goldberg, S.L. Fine, Airlie House, Warrenton, VA, PHS pub. no. 1890, 691–700 (1969).

"Measurement of Cross-sections with Neutrons as Targets, II," M. Widgoff, A.M. Cormack, A.M. Koehler, The Physical Review, 177 (no. 5, part 1), 2016–2022 (1969).

"Pituitary Suppression in Diabetic Retinopathy by Proton Beam in Surgically 'Unfit' Patients," R.N. Kjellberg, J.W. McMeel, N.L. McManus, A.M. Koehler, in *Symposium on the Treatment of Diabetic Retinopathy*, ed. M.F. Goldberg, S.L. Fine, Airlie House, Warrenton, VA, PHS pub. no. 1890:249–276 (1969).

"Stereotactic Proton Hypophysectomy for Hyperpituitarism: Acromegaly,100 Cases; Cushing's Disease, 22 Cases," R.N. Kjellberg, B. Kliman, Abstract in Excerpta Medical Int. Cong. Series, 193:31 (1969).

"A Thin Parallel-walled Liquid Hydrogen Target," G.M. Polucci, A.M. Koehler, J.N. Palmieri, Nuc. Inst. and Meth., 71:218–220 (1969).

"The Measurement of the R Parameter for Proton-Deuteron Elastic Scattering and for Quasi-Free Proton-Proton and Proton -Neutron Scattering in Deuterium," Jacques Lefrancois, Ph.D. thesis (1969).

1970

"The neutron-proton interaction below 30 MeV," in Proceedings of the European-American Nuclear Data Committee Symposium on Neutron Standards and Flux Normalization, Argonne National Laboratory (1970).

"The Bragg Peak of a 160 MeV Proton Beam [Abstract]," A.M. Koehler, Phys. Med. Biol., 15:143 (1970).

"Perturbed Angular Correlation Measurement on Rh^{100} in a Ni Host: Critical

Exponent β for Ni," R. Reno, C. Hohenemser, Physical Review Letters, 25(15):1007–1011 (1970).

"Studies of the Cellular Radiosensitivity of Transplants of Murine Ependymomas Irradiated in Vivo," K. Stratton, A. Anderson, Int. J. Rad. Biol., 18(1):1–23 (1970).

1971

"Lifetime and g-factor of the 74.8 KeV State of Rh[100]," R. Reno, M. Fishbein, C. Hohensemser, Nuclear Physics, A163:161–165 (1971).

"Proton beam therapy, R.N. Kjellberg, B. Kliman, Letter to the editor, New England Journal of Medicine, 284:333 (Feb. 1971).

"Proton Radiographic Techniques," A.M. Koehler, Trans. Am. Nucl. Soc., 14(1): 95 (1971).

"Therapy of Acromegaly," B. Kliman, R.N. Kjellberg, Letter to the editor, New England Journal of Medicine, 284:673 (Mar. 1971).

"Use of Protons for Radiotherapy" A.M. Koehler, Proc. of the Symposium on Pion and Proton Radiotherapy, Nat. Accelerator Lab., (Dec. 1971).

1972

"The Bragg Peak Proton Beam in Noninvasive Hypophysectomy," R.N. Kjellberg, Hospital Practice, 7(10):95–102 (1972).

"Le Bragg Peak Protonique en Neuro-chirurgie Stereotaxique," R.N. Kjellberg, N.C. Nguyen, B. Kliman, Neuro-chirurgie, 18(3):235-265 (1972).

"Le Bragg Peak Des Protons en Neuro-chirurgie Stereotaxique," R.N. Kjellberg, Sandoz Editions de Neurologie et Psychiatrie, 3–11 (Apr. 1972).

"Critical Fluctuations in Nickel Observed by Time Differential Perturbed Angular Correlations on Rh[100] Impurity Nuclei," C. Hohenemser, R. Reno, in AIP Conference Proceedings: Magnetism and Magnetic Materials 1971, ed. C. Graham, J. Rhyne, 1256–1260 (1972).

"Neuropathologic Effects of Proton-beam Irradiation in Man: I. Dose Response Relationships after Treatment of Intracranial Neoplasms," S.L. Nielsen, R.N. Kjellberg, A.K. Asbury, A.M. Koehler, Acta Neuropathologica, 20:348–356 (1972).

"Neuropathologic Effects of Proton-beam Irradiation in Man: II. Evaluation After Pituitary Irradiation, S.L. Nielsen, R.N. Kjellberg, A.K. Asbury, A.M. Koehler, Acta Neuropathologica, 21:76–82 (1972).

Protons in Radiation Therapy: Comparative Dose Distributions for Protons, Photons and Electrons, A.M. Koehler, W.M. Preston, Radiology, 104(1): 191–195 (1972).

Proton Radiation Therapy: Existing Technology," A.M. Koehler, in *Potentialities of Proton Radiotherapy* (report of symposium Aug. 1972), ed. J. Archambeau, BNL 50365 (1972).

"Radiation Therapy," R.N. Kjellberg, letter to the editor, Science, 176, 1071 (June 1972).

"Medical Treatment and Diagnosis Using 160 MeV Protons," A.M. Koehler, presented at the Sixth Int. Cyclotron Conference, Vancouver (1972). In *Cyclotrons 1972*, AIP Conf. Proc., 9, ed. J. Burgerjon, A. Strathdee, 586–600 (1972).

"Pituitary-adrenal Function after Proton Beam Therapy for Cushing's Disease," B. Kliman, R.N. Kjellberg, presented at Fourth Int. Congress of Endocrinology, Washington, DC (June 1972). Abstract in Excerpta Medica Int. Cong. Series, 256, 84 (1972).

"Proton Radiography as a Diagnostic Tool," V.W. Steward, A.M. Koehler, presented at AAPM Winter Meeting, Chicago (Nov. 1972). Abstract in AAPM Bulletin, 6(4):194 (1972).

1973
"Applications of Proton Radiation to Cancer Therapy," A.M. Koehler, Proc. of the Grantees Conference on Instrumentation Technology, NSF, Washington, DC (Sept. 1973).

"In Vivo Calcium Determination by Proton Activation Analysis" [Extended abstract], R. Eilbert, Int. Conf. on Bone Mineral Measurement, Chicago, IL, 151–154 (Oct. 1973).

"Proton Beam Radiography in Tumor Detection," V.W. Steward, A.M. Koehler, Science, 173:913–914 (1973).

Proton radiographic detection of strokes, V.W. Steward, A.M. Koehler, Nature, 245:38–40 (1973).

"Proton Radiography," A.M. Koehler, H. Berger, in Research Techniques in Non-destructive Testing, vol. 2, ed. R.S. Sharp, Academic Press, London-New York, 1–30 (1973).

"A System for Therapy of Pituitary Tumors," R.N. Kjellberg, B. Kliman, in Diagnosis and Treatment of Pituitary Tumors, (proc. of conf. Bethesda, MD, Jan. 1973), ed. P. Koehler, G. Ross, Excepta Medica, Int. Cong. Series, 303:234–252 (1973).

1974
"Bragg Peak Proton Treatment for Pituitary-related Conditions," R.N. Kjellberg, B. Kliman, Proc. of the Royal Society of Medicine, 67:32–33 (1974).

"Dose-limiting Tissues in Relation to Type and Location of Tumors: Implications for Efforts to Improve Radiation Dose Distributions," H.D. Suit, M. Goitein, European J. of Cancer, 10:217–224 (1974).

"Experimental Ocular Irradiation with Protons," I.J. Constable, A.M. Koehler, Investigative Ophthalmology, 13:280–287 (1974).

"Harvard's Cyclotron," A.M. Koehler, Letters to the editor, Harvard Magazine, 76(8):4 (1974).

"Non-invasive Hypophysectomy for Acromegaly by the Bragg Peak Proton Technique," R.N. Kjellberg, in Recent Progress in Neurological Surgery (Proc. of the Symposia of the Fifth Int. Cong. of Neurological Surgery, Tokyo 1973), Excerpta Medica, Amsterdam, 99–109 (1974).

"Proton Radiography in the Diagnosis of Breast Carcinoma," V.W. Steward, A.M. Koehler, Radiology, 110(1):217–221 (1974).

"Proton Radiography of a Human Brain Tumor within the Skull: A Preliminary Report," Surgical Neurology, 2(4):283–284 (1974).

"Sensitivities of Some Photographic Films and Screens for Proton Radiography," R.F. Eilbert, A.M. Koehler, Nuc. Inst. and Meth., 114:581–585 (1974).

1975

"Bragg Peak Proton Hypophysectomy for Hyperpituitarism, Induced Hyperpituitarism and Neoplasms," R.N. Kjellberg, B. Kliman, in Progress in Neurological Surgery, 6, ed. H. Krayenbuhl, P. Maspes, W. Sweet, S. Karger, Basel, 295–325 (1975).

"Current status of Proton Therapy in North America," H.D. Suit, M. Goitein, J. Tepper, A.M. Koehler, E. Gragoudas, Proc. of Int. Workshop in Particle Radiation Therapy, Key Biscayne, FL, 529–540 (1975).

"Cyclotron Produced Rubidium 81 For Nuclear Medicine," R.J. Schneider, C.J. Goldberg, J.D. Idoine, B.L. Holman, A.G. Jones, Medical Physics, 2:148 (1975). [Abstract].

"Exploratory Study of Proton Radiation Therapy Using Large Field Techniques and Fractionated Dose Schedules," H.D. Suit, M. Goitein, J. Tepper, A.M. Koehler, R.A. Schmidt, R.J. Schneider, Cancer, 35(6):1646–1657 (1975).

"The Influence of Tissue Inhomogeneities on the Dose Distribution of Charged Particle Beams," M. Goitein, H.D. Suit, Proc. of Int. Workshop in Particle Radiation Therapy, Key Biscayne, FL, 138–149 (1975).

"Radiobiological Studies of a High Energy Modulated Proton Beam Utilizing Cultured Mammalian Cells," J. Robertson, J. Williams, R. Schmidt, J. Little, D. Flynn, H.D. Suit, Cancer, 35(6):1664–1677 (1975).

"Range Modulators for Protons and Heavy Ions," A.M. Koehler, R.J. Schneider, J.M. Sisterson, Nuc. Instr. and Meth., 131:437–440 (1975).

"A System of Therapy of Pituitary Tumors – Bragg Peak Proton Hypophysectomy," R.N. Kjellberg, in Tumors of the Nervous System, ed. H. G. Seydel, John Wiley and Sons, New York, 145–174 (1975).

"Treatment of Acromegaly," R.N. Kjellberg, B. Kliman, Letter to the Editor, The Lancet, 630–631 (15 Mar. 1975).

"Dose Perturbation of Charged Particle Beams by Discontinuities – Theoretical Analysis," M. Goitein, A.M. Koehler, R.J. Schneider, J.Y. Ting, presented at the 17th Annual Meeting of the AAPM, San Antonio (Aug. 1975). Abstract, Medical Physics, 2:154 (1975).

"System of Therapy in 337 Procedures for Acromegaly Emphasizing Bragg Peak Proton Hypophysectomy," R.N. Kjellberg, B. Kliman, presented at Int. Symposium on Growth Hormone and Related Peptides, Universit° degli Studi di Milano (1975). [Abstract].

1976

"Measurements of Proton Stopping Power and Density on Fresh Human Tissues," A.M. Koehler, K.N. Johnson, Technical report of work performed with the Univ. of Chicago under NCI grant (July 1976).

"Production of Rb^{81} by the reaction $Rb^{85}(p,Sn)Sr^{81}$ and decay of SR^{81}," R.J. Schneider, C.J. Goldberg, Int. J. of Appl. Rad. and Isotopes, 27:189-190 (1976).

"Quantitative Proton Tomography: Preliminary Experiments," A.M. Cormack, A.M. Koehler, Phys. Med. Biol., 21(4):560–569 (1976).

"A range Modulator to Produce Uniform 38k Yield," R.F. Eilbert, A.M. Koehler, J.M. Sisterson, Int. J. of Appl. Rad. and Isotopes, 27:707–711 (1976).

"Small-field Irradiation of Monkey Eyes with Protons and Photons," I.J. Constable, M. Goitein, A.M. Koehler, R.A. Schmidt, Radiation Research, 65: 304–314 (1976).

"Treatment of Acromegaly by Proton Hypophysectomy," R.N. Kjellberg, B. Kliman, in Controversy in Surgery, ed. R. Varco, J. Delaney, W. Saunders Co., Toronto, 392–405 (1976).

1977

"Clinical Experience and Expectations with Protons and Heavy Ions," H.D. Suit, M. Goitein, J.E. Tepper, L. Verhey, A.M. Koehler, R.J. Schneider, E. Gragoudas, Int. J. of Rad. Onc., Biol., Phys., 3:115–125 (1977).

"Flattening of Proton Dose Distributions for Large-field Radiotherapy," A.M. Koehler, R.J. Schneider, J.M. Sisterson, Medical Physics, 4:297–301 (1977).

"In Vivo Determination of Rbe in a High Energy Modulated Proton Beam Using Normal Tissue Reactions and Fractionated Dose Schedules," J. Tepper, L. Verhey, M. Goitein, H.D. Suit, A.M. Koehler, Int. J. of Rad. Onc., Biol., Phys., 2:1115–1122 (1977).

"The Measurement of Tissue Heterodensity to Guide Charged Particle Radiotherapy," M. Goitein, Int. J. of Rad. Onc., Biol., Phys., 3:27–33 (1977).

"Partial Body Calcium Determination in Bone by Proton Activation Analysis," R.F. Eilbert, J.M. Sisterson, R. Wilson, S.J. Adelstein, Phys. Med. Biol., 22:817–830 (1977).

"Proton Irradiation of Small Choroidal Malignant Melanomas," E. Gragoudas, M. Goitein, A.M. Koehler, L. Verhey, J. Tepper, H.D. Suit, R. Brockhurst, I.J. Constable, Am. J. Ophthalmology, 83:665–673 (1977).

"Proton Penetration and Control in Nonhomogeneous Phantoms," C.L. Wingate, J.O. Archambeau, A.M. Koehler, G.W. Bennett, Medical Physics, 4(3):198–201 (1977).

"Quantification of Flow in a Dynamic Phantom using Rb^{81}-81mKr, and a NaI Detector," J.D. Idoine, B.L. Holman, A.G. Jones, R.J. Schneider, K.L. Schroeder, R.E. Zimmerman, J. of Nuclear Medicine, 18:570–578 (1977).

"Bragg peak Proton Radiosurgery for Arteriovenous Malformation of the Brain," R.N. Kjellberg, presented at First Int. Seminar on the Use of Proton Beams in Radiation Therapy, Moskow (Dec. 1977). See also Proc. (1979).

"Bragg Peak Proton Radiosurgical Hypophysectomy for Pituitary Adenomas," R.N. Kjellberg, presented at First Int. Seminar on the Use of Proton Beams in Radiation Therapy, Moskow (Dec. 1977).

"Calorimetric Dosimetry for Cyclotron Produced Neutron and Proton Fields," J. McDonald, I. Ma, J. Laughlin, F. Attix, L. August, M. Goitein, L. Verhey, A.M. Koehler, presented at 25th Annual Meeting of the Radiation Research Society, San Juan, PR (May 1977). [Abstract].

"Control of Proton Dose Distributions for Large Field Therapy," R. Schneider, A.M. Koehler, J. Sisterson, R. Schmidt, presented at Fourth Int. Conf. on Medical Physics, Ottawa (July 1977). Abstract, Phys. Med. Biol., 22:156 (1977).

"Curative Radiation Therapy Employing Proton Beam Techniques," H.D. Suit, M. Goitein, J. Tepper, O. Mendiondo, L. Verhey, A.M. Koehler, C. Friedberg, R. Sedlacek, E. Gragoudas, presented at First Int. Seminar on the Use of Proton Beams in Radiation Therapy, Moskow (Dec. 1977).

"Facilities for Large Field Proton Therapy at the Harvard Cyclotron Laboratory," A.M. Koehler, R.J. Schneider, J.M. Sisterson, M. Wagner, C. Friedberg, M. Goitein, L. Verhey, presented at First Int. Seminar on the Use of Proton Beams in Radiation Therapy, Moskow (Dec. 1977).

"The Influence of Inhomogeneities on Charged Particle Beams – Slivers," M. Goitein, J. M. Sisterson, presented at the Fourth International Conference on Medical Physics, Ottawa (July 1976). Abstract, Physics in Canada, 32 (July 1976) and Phys. Med. Biol., 22, 156 (1977).

"Proton Beam Irradiation of Ocular Melanomas with Submillimetrer Precision," A.M. Koehler, K. Johnson, R.A. Schmidt, R.J. Schneider, M. Wagner, I.J. Constable, M. Goitein, H.D. Suit, J. Tepper, L. Verhey, presented at the Fourth International Conference on Medical Physics, Ottawa (July 1976). Abstract, Physics in Canada, 32 (July 1976) and Phys. Med. Biol., 22:156 (1977).

"Radiobiological Evaluation of a Modulated Energy 160 Mv Proton Beam," H.D. Suit, M. Goitein, J. Tepper, A.M. Koehler, L. Verhey, C. Friedberg, R. Schneider, presented at First Int. Seminar on the Use of Proton Beams in Radiation Therapy, Moskow (Dec. 1977).

"Use of Proton Beams for Treatment of Choroidal Melanomas," E. Gragoudas, M. Goitein, A.M. Koehler, M. Wagner, L. Verhey, J. Tepper, H. Suit, R. Schneider, K. Johnson, presented at First Int. Seminar on the Use of Proton Beams in Radiation Therapy, Moskow (Dec. 1977).

1978

"Bragg Peak Proton Beam Treatment of Arteriovenous Malformations of the Brain," R.N. Kjellberg, C.E. Poletti, G.H. Roberson, R.D. Adams, in Neurological Surgery, Int. Congress Series No. 433, Proceedings of the Sixth Int. Congress of Neurological Surgery, 181–187 (1978).

"Compensation for Inhomogeneities in Charged Particle Radiotherapy Using Computed Tomography," M. Goitein, Int. J. Rad. Onc., Biol., Phys., 4: 499–508 (1978).

"$Ca^{40}(p,2pn)$ K^{38} Total Nuclear Cross Section," J.M. Sisterson, A.M. Koehler, R.F. Eilbert, Physical Review C, 18(1):582–583 (July 1978).

"A Heavy Particle Comparative Study – Part I: Depth-dose Distributions," M.R. Raju, H.I. Amols, J.F. Dicello, J. Howard, J.T. Lyman, A.M. Koehler, R. Graves, J.B. Smathers, British J. of Radiology, 51:699–703 (1978).

"The Influence of Thick Inhomogeneities on Charged Particle Beam," M. Goitein, J.M. Sisterson, Radiation Research, 74:217–230 (1978).

"Measurements and Calculations of the Influence of Thin Inhomogeneities on Charged Particle Beams," M. Goitein, G.T.Y. Chen, J.Y. Ting, R.J. Schneider, J.M. Sisterson, Medical Physics, 5(4):265–273 (July–Aug. 1978).

"Proton Irradiation of Choroidal Melanomas: Preliminary Results," E. Gragoudas, M. Goitein, A.M. Koehler, I.J. Constable, M. Wagner, L. Verhey, J. Tepper, H.D. Suit, R.J. Brockhurst, R.J. Schneider, K. Johnson, Archives of Ophthalmology, 96:1583–1591 (1978).

"A Technique for Calculating the Influence of Thin Inhomogeneities on Charged Particle Beams," M. Goitein, Medical Physics, 5(4):258–264 (July–Aug. 1978).

"Clinical Experience with the 160 MeV Proton Beam and Some Implications for Designers," A.M. Koehler, K. Johnson, presented at Eighth Int. Conference on Cyclotrons and Their Applications, ICUF, Bloomington, IN (Sept. 1978).

"Compensation Techniques in Charged Particle Therapy," L. Verhey, M. Goitein, J. Sisterson, presented at the 20th Annual Meeting of the AAPM, San Francisco (July 1978). Abstract, Medical Physics, 5:325 (1978).

1979

"Bragg Peak Proton Radiosurgery for Arteriovenous Malformations of the Brain," R.N. Kjellberg, Proc. First Int. Sem. on the Uses of Proton Beams in Radiation Therapy, Moskow 1977, 3:12–21, Moskow (1979).

"Bragg Peak Proton Radiosurgical Hypophysectomy for Pituitary Adenomas," R.N. Kjellberg, Proc. First Int. Sem. on the Uses of Proton Beams in Radiation Therapy, Moskow 1977, 3:22–34, Moskow (1979).

"Clinical Experience with the 160 MeV Proton Beam and Some Implications for Designers," A.M. Koehler, K. Johnson, IEEE Transactions on Nuclear Science, NS-26(2):2253–2256 (April 1979).

"Computed Tomography in Planning Radiation Therapy," M. Goitein, Int. J. of Rad. Onc., Biol., Phys., 5:445–447 (1979).

"Curative Radiation Therapy Employing Proton Beam Techniques," H.D. Suit, M. Goitein, J. Tepper, O. Mendiondo, L. Verhey, A.M. Koehler, C. Friedberg, R. Sedlacek, E. Gragoudas, Proc. First Int. Sem. on the Uses of Proton Beams in Radiation Therapy, Moskow 1977, 3:77–87, Moskow (1979).

"The Determination of Absorbed Dose in a Proton Beam for Purposes of Charged Particle Radiation Therapy," L.J. Verhey, A.M. Koehler, J.C. McDonald, M. Goitein, I. Ma, R.J. Schneider, M. Wagner, Radiation Research, 79:34–54 (1979).

"Facilities for Large Field Proton Therapy at the Harvard Cyclotron Laboratory," A.M. Koehler, R.J. Schneider, J.M. Sisterson, M. Wagner, C. Friedberg, M. Goitein, L. Verhey, Proc. First Int. Sem. on the Uses of Proton Beams in Radiation Therapy, Moskow 1977, 2:104–120, Moskow (1979).

"The influence of Inhomogeneities upon the Dose Distribution of Protons," M. Goitein, Proc. First Int. Sem. on the Uses of Proton Beams in Radiation Therapy, Moskow 1977, 1:95–105, Moskow (1979).

"Isoeffective Dose Parameters for Brain Necroses in Relation to Proton Radiosurgical Dosimetry," R.N. Kjellberg, in Stereotactic Cerebral Radiation, ed. G. Szikla, Elsevier Press, Amsterdam, 157–166 (1979).

"Lifetime Effectiveness – A System of Therapy for Pituitary Adenomas, Emphasizing Bragg Peak Proton Hypophysectomy," R.N. Kjellberg, B. Kliman, in Recent Advances in the Diagnosis and Treatment of Pituitary Tumors, ed. J.A. Linfoot, Raven Press, New York, 269–288 (1979).

"Long-term Observation of Proton Irradiated Monkey Eyes," E. Gragoudas, N.Z. Zakov, D.M. Albert, I.J. Constable, Arch. Ophthalmology, 97:2184–2191 (Nov. 1979).

"Multiple Choroidal Metastasis from Bronchial Carcinoid Treated with Photocoagulation and Proton Beam Irradiation," E. Gragoudas, J. Carrol, Am. J. of Ophthalmology, 87(3):299–304 (1979).

"An Overview of Cyclotron Treatment, Bragg Peak Proton Hypophysectomy and Bragg Peak Radiosurgery for Arteriovenous Malformation of the Brain," J. D'Agostino, L. Pelczynski, J. of Neurosurgical Nursing, 11(4): 208–214 (1979).

"Proton Dosimetry at Harvard Cyclotron Laboratory," M. Goitein, L. Verhey, Proc. First Int. Sem. on the Uses of Proton Beams in Radiation Therapy,

Moskow 1977, 1:201–207, Moskow (1979).

"Proton Irradiation of Malignant Melanoma of the Ciliary Body," E. Gragoudas, M. Goitein, A.M. Koehler, M. Wagner, L. Verhey, J. Tepper, H.D. Suit, R.J. Schneider, K.N. Johnson, British J. of Ophthalmology, 63(2): 135–139 (Feb. 1979).

"Proton Radiation as Boost Therapy for Localized Prostatic Carcinoma," W.U. Shipley, J.E. Tepper, G.R. Prout, L.J. Verhey, O.A. Mendiondo, M. Goitein, A.M. Koehler, H.D. Suit, J.A.M.A., 241:1912–1915 (May 1979).

"Proton Radiosurgery for Functioning Pituitary Adenoma," R.N. Kjellberg, B. Kliman, in Clinical Management of Pituitary Disorders, ed. G.T. Tindall, W.F. Collins, Raven Press, New York, 315–334 (1979).

"Radiobiological Evaluation of a Modulated Energy 160 Mv Proton Beam," H.D. Suit, M. Goitein, J. Tepper, A.M. Koehler, L. Verhey, C. Friedberg, R. Schneider, Proc. First Int. Sem. on the Uses of Proton Beams in Radiation Therapy, Moskow 1977, 2:52–61, Moskow (1979).

"Radiobiology, Dosimetry and Stereotactic Technique for Bragg Peak Proton Radiosurgical Procedures," R.N. Kjellberg, Proc. First Int. Sem. on the Uses of Proton Beams in Radiation Therapy, Moskow 1977, 2:69–83, Moskow (1979).

"Rx: The Cyclotron," R. Wilson (also: "How It Works," K. Johnson, A. Coggeshall), Harvard Magazine, 82(2):58–62 (1979).

"Soft Errors Induced by Energetic Protons," R.C. Wyatt, P.J. McNulty, P. Toumbas, P.L. Rothwell, R.C. Filz, IEEE Trans. on Nuclear Science, NS-26(6):4905–4910 (Dec. 1979).

"Stereotactic Bragg Peak Proton Radiosurgery Method," R.N. Kjellberg, in Stereotactic Cerebral Radiation, ed. G. Szikla, Elsevier Press, Amsterdam, 93–100 (1979).

"Stereotactic Bragg Peak Proton Radiosurgery Results," R.N. Kjellberg, in Stereotactic Cerebral Radiation, ed. G. Szikla, Elsevier Press, Amsterdam, 233–240 (1979).

"Therapy of Recurrent Acromegaly," R.N. Kjellberg, Neurosurgery, 4(6):573–574 (1979).

"Use of Proton Beams for Treatment of Choroidal Melanomas," E. Gragoudas, M. Goitein, A.M. Koehler, M. Wagner, L. Verhey, J. Tepper, H. Suit, R. Schneider, K. Johnson, Proc. First Int. Sem. on the Uses of Proton Beams in Radiation Therapy, Moskow 1977, 3:63–75, Moskow (1979).

"Biological Properties of Protons," H.D. Suit, M. Urano, M. Goitein, L. Verhey, presented at 6th Int. Cong. of Radiation Research, Tokyo (1979), also in Proc., ed. S. Okada, M. Imamura, T. Terashima, H. Yamaguchi, Toppan Printing, Tokyo, 771–779 (1979).

"Proton Activation Analysis for Elemental Analysis," R.J. Schneider, J.M. Sisterson, NES/APS Meeting, Bates College (Sept. 1979). [Abstract].

1980

"Accelerated Protons for Biology and Medicine," R.J. Schneider, Trans. A. Nucl. Soc., 35:25–26 (1980).

"Clinical Experience with Proton Beam Radiation Therapy," H.D. Suit, M. Goitein, J.E. Munzenrider, L. Verhey, E. Gragoudas, A.M. Koehler, M. Urano, W.U. Shipley, R.M.Linggood, C. Friedberg, M. Wagner, J. of the Canadian Assoc. of Radiologists, 31:35–39 (March 1980).

"Proton Upsets in LSI Memories in Space," P. McNulty, R.Wyatt, G. Farrell, R. Filz, P. Rothwell, in Space Systems and Their Interactions with the Earth's Space Environment, ed. H. Garrett, C. Pike, American Institute of Aeronautics and Astronautics, New York, 413–433 (1980).

"Radiosurgery Therapy for Pituitary Adenoma," R.N. Kjellberg, B. Kliman, in The Pituitary Adenoma, ed. K.D. Post, Plenum Publishing Corp., 459–478 (1980).

"Rationale for Use of Charged-particle and Fast-neutron Beams in Radiation Therapy," H. Suit, M. Goitein, in Rad. Biol. in Cancer Research, ed. R. Meyn, R. Withers, Raven Press, New York, 547–565 (1980).

"Proton Beam Irradiation: An Alternative to Enucleation for Intraocular Melanomas," E. Gragoudas, M. Goitein, L. Verhey, J. Munzenrider, H.D. Suit, A.M. Koehler, Ophthalmology, 87:571–581 (June 1980).

"Radiosurgery for Pituitary Adenoma with Bragg Peak Proton Beam," R.N. Kjellberg, B. Kliman, B.J. Swisher, in Pituitary Adenomas, ed. P.Derone, C. Jedynak, F. Peillon, Asclepios Publishers, 209–217 (1980).

"Radiosurgery Therapy for Pituitary Adenoma," R.N. Kjellberg, B. Kliman, in The Pituitary Adenoma, ed. K.D. Post, Plenum Publishing Corp., 459–478 (1980).

"Rationale for use of Charged-particle and Fast-neutron Beams in Radiation Therapy," H. Suit, M. Goitein, in Rad. Biol. in Cancer Research, ed. R. Meyn, R. Withers, Raven Press, New York, 547–565 (1980).

"Relative Biological Effectiveness of a High Energy Modulated Proton Beam Using a Spontaneous Murine Tumor in Vivo," M. Urano, M. Goitein, L. Verhey, O. Mendiondo, H.D. Suit, A.M. Koehler, Int. J. Rad. Onc., Biol, Phys., 6:1187–1193 (1980).

"Upset Phenomena Induced by Energetic Protons and Electrons," P. McNulty, G. Farrell, R. Wyatt, P. Rothwell, C. Filz, J. Bradford, IEEE Transactions on Nuclear Science, NS-27(6):1516–1522 (1980).

1981

"Future Prospects of Radiation Therapy with Protons," J. Munzenrider, W.U. Shipley, L. Verhey, in Seminars in Oncology, 8(1):110–124 (1981).

"Proton-induced Nuclear Reactions in Silicon," P. McNulty, G. Farrell, W. Tucker, IEEE Transactions in Nuclear Science, NS-28(6):4007–4012 (1981).

"Treating Tumors with the Harvard Cyclotron," A. Coggeshall, K. Johnson, Encyclopaedia Britannica Medical and Health Annual, 219–222 (1981).

"Automated Range Compensation for Proton Therapy," M. Wagner, presented at the meeting of The Radiological Society of North America, Chicago, IL (Nov. 1981).

"Calcium/phosphorus Ratio Determination by High-energy Proton Activation Analysis," J.M. Sisterson, R.J. Schneider, P. Tibbetts, M. Grynpas, L.C. Bonar, presented at the 6th Conference on Modern Trends in Activation Analysis, Toronto (June 1981).

"Cost-effective Proton Beam Therapy with a Cyclotron," A.M. Koehler, presented at the 9th Int. Conference on Cyclotrons and their Applications, Caen, France (Sept. 1981). See also Proc. (1982).

"Dosimetry Intercomparisons between Heavy Charged Particle Radiotherapy Facilities," A. Smith, R. Hilco, J. DiCello, P. Fessenden, M. Henkelman, K. Hogstrom, G. Lam, J. Lyman, H. Blattmann, M. Salzmann, D. Reading, L. Verhey, presented at Int. Workshop on Pion and Heavy Ion Radiotherapy: Pre-clinical and Clinical Studies, Vancouver (July 1981). See also Proc. (1983).

"Problems of Inhomogeneities in Particle Beam Therapy," L. Verhey, M. Goitein, presented at Int. Workshop on Pion and Heavy Ion Radiotherapy: Pre-clinical and Clinical Studies, Vancouver (July 1981). See also Proc. (1983).

"Proton Therapy at Harvard," J. Munzenrider, presented at Int. Workshop on Pion and Heavy Ion Radiotherapy: Pre-clinical and Clinical Studies, Vancouver (July 1981). See also Proc. (1983).

"Whole-body calcium measurements in animals using proton activation analysis," A.M. Koehler, A.J. Zoesman, R.I. Abrams, R.N. Neer, 6th Conference on Modern Trends in Activation Analysis, Toronto (June 1981).

1982

"Automated Range Compensation for Proton Therapy," M.S. Wagner, Med. Phys., 9(5):749–752 (Sept.–Oct. 1982).

"Cost-effective Proton Beam Therapy with a Cyclotron," A.M. Koehler, in Proc. of the 9th Int. Conference on Cyclotrons and Their Applications, Caen, France,Sept. 1981, Editions de Physique, Paris, 667–672 (1982).

"Definitive Radiation Therapy for Chordoma and Chondrosarcoma of Base of Skull and Cervical Spine," H.D. Suit, M. Goitein, J. Munzenrider, L. Verhey, K. Davis, A.M. Koehler, R. Linggood, R. Ojemann, Journal of Neurosurgery, 56:377–385 (1982).

"Evaluation of the Clinical Applicability of Proton Beams in Definitive Fractionated Radiation Therapy," H.D. Suit, M. Goitein, J. Munzenrider, L. Verhey, P. Blitzer, E. Gragoudas, A.M. Koehler, M. Urie, R. Gentry, W.

Shipley, M. Urano, J. Duttenhaver, M. Wagner, Int. J. Rad. Onc., Biol., Phys., 8(12):2199–2205 (1982).

"Microdosimetric Aspects of Proton-induced Nuclear Reactions in Thin Layers of Silicon," G.E. Farrell, P.J. McNulty, IEEE Trans. on Nuclear Science, NS-29:2012–2016 (1982).

"Planning Treatment with Heavy Charged Particles," M. Goitein, M. Abrams, R. Gentry, M. Urie, L. Verhey, M. Wagner, Int. J. Rad. Onc., Biol., Phys., 8: 2065–2070 (1982).

"Precise Prositioning of Patients for Radiation Therapy," L. Verhey, M. Goitein, P. McNulty, J. Munzenrider, H.D. Suit, Int. J. Radiation Onc., Biol., Phys., 8:289–294 (1982).

"Proton beam Irradiation of Uveal Melanomas: Results of a 5½ Year Study," E. Gragoudas, M. Goitein, L. Verhey, J. Munzenrider, M. Urie, H.D. Suit, A.M. Koehler, Archives of Ophthalmology, 100:928–934 (1982).

"Proton Beam Therapy," L. Verhey, J. Munzenrider, Annual Review Biophys., Bioeng., 11:331–357 (1982).

"Proton Radiation as Boost Therapy for Patients with Locally Advanced Prostatic Adenocarcinoma: A Comparison with Megavoltage X-rays Alone," W.U. Shipley, J.R. Duttenhaver, L. Verhey, M. Goitein, J. Munzenrider, T. Perrone, P. McNulty, G. Prout, H.D. Suit, Proc. of Int. Symposium on the Management of Prostatic Carcinoma, Genova, 93–102 (Jan. 1982).

"Proton Therapy," M. Goitein, P. Blitzer, J. Duttenhaver, R. Gentry, B. Gottschalk, E. Gragoudas, K. Johnson, A.M. Koehler, J. Munzenrider, W. Shipley, H.D. Suit, M. Urano, M. Urie, L. Verhey, M. Wagner, Proc. Int. Conf. on Applications of Physics to Medicine and Biology, Trieste, World Scientific Publishing Co., Singapore, 27–44 (1982).

"The reaction $P^{31}(p,3p)Al^{29}$ and its Use in Measuring the Ca/P Molar Ratio in Chemical Compounds," J.M. Sisterson, R.J. Schneider, P. Tibbetts, M. Grynpas, L.C. Bonar, J. Radioanal. Chem., 71:509–518) (1982).

"The 733 as a Low Input-impedance Preamplifier for Current-division Use," B. Gottschalk, Nucl. Instr. and Meth., 196:447–448 (1982).

"Single Particle Effects in Microelectronics; A Problem in Microdosimetry," P. McNulty, in Radiation Protection: Eighth Int. Symposium on Microdosimetry, eds. J. Booz, H. Ebert, Commission of the European Communities, Luxemborg, 921–933 (1982).

"Beam Control and Treatment Facilities for Large-field Proton Radiotherapy," A.M. Koehler, Karl Bremer Hospital, Cape Town, South Africa (March 1982).

"Bone Calcium Measured by Proton Activation," A.M. Koehler, A.J. Zoesman, R.M. Neer, presented at Annual Conference of South African Association of Physicists in Medicine and Biology, Durban (March 1982).

"Design of an Accelerator for Therapy with Proton Beams," B. Gottschalk,

World Congress on Medical Physics and Biomedical engineering, Hamburg (Sept. 1982).

"Proton Beam Irradiation of Uveal Melanomas: An Alternative to Enucleation," E. Gragoudas, J. Sisterson, K. Johnson, XXIV Int. Cong. of Ophthalmology, San Francisco (Oct. 1982).

"Proton Beam Therapy at the Harvard Cyclotron Laboratory," A.M. Koehler, presented at Annual Conference of South African Association of Physicists in Medicine and Biology, Durban (March 1982).

"Radiotherapy with 160 MeV Protons," B. Gottschalk, A.M. Koehler, R. Wilson, Presented at the Int. Symposium on Applications and Technology of Ionizing Radiations, Riyadh, Saudi Arabia (March 1982). See also Proc. (1983).

"Small-field Proton Beam Irradiations of Intracranial and Intraocular Targets," A.M. Koehler, Groote Schuur Hospital, Cape Town, South Africa (March 1982).

1983
"Beam Scanning for Heavy Charged Particle Radiotherapy," M. Goitein, G.T.Y. Chen, Medical Physics, 10(6):831–840 (Nov./Dec. 1983).

"Bragg-peak Proton Beam Therapy for Arteriovenous Malformations of the Brain," R.N. Kjellberg, T. Hanamura, K.R. Davis, S.L. Lyons, R.D. Adams, New England J. of Medicine, 309:269–274 (1983)

"Charge-balancing Current Integrator with Large Dynamic Range," B. Gottschalk, Nucl. Instr. & Meth., 207:417–421 (1983).

"Charge-deposition Spectra in Thin Slabs of Silicon Induced by Energetic Protons," S. El Teleaty, G.E. Farrell, P.J. McNulty, IEEE Trans. on Nuclear Science, NS-30(6):4394–4397 (1983).

"Charged Particles Cause Microelectronics Malfunction in Space," P. McNulty, Physics Today, 36, 9:108–109 (1983).

"Ciliary Body and Choroidal Melanomas Treated by Proton Beam Irradiation: Histopathologic Study of Eyes," J. Seddon, E. Gragoudas, D. Albert, Archives of Ophthalmology, 101:1402–1408 (Sept. 1983).

"Compensating for Heterogeneities in Proton Radiation Therapy," M. Urie, M. Goitein, M. Wagner, Phys. Med. Biol., 29(5):553–566 (1983).

"Determination of the Radioprotective Effects of Topical Applications of Mea, Wr-2721, and N-acetylcysteine on Murine Skin," L. Verhey, R. Sedlacek, Radiation Research, 93:175–183 (1983).

"Dosimetry Intercomparisons between Heavy Charged Particle Radiotherapy Facilities," A. Smith, R. Hilco, J. DiCello, P. Fessenden, M. Henkelman, K. Hogstrom, G. Lam, J. Lyman, H. Blattmann, M. Salzmann, D. Reading, L. Verhey, in Pion and Heavy Ion Radiotherapy: Pre-clinical and Clinical Studies, ed. L. Skarsgard, Elsevier Biomedical, New York-Amsterdam-Oxford, 49–61 (1983).

"Energy of Proton Accelerator Necessary for Treatment of Choroidal Melanomas," M. Goitein, R. Gentry, A.M. Koehler, Int. J. Rad. Onc., Biol., Phys., 9:259–260 (1983).

"Investigation of Buildup Dose from Electron Contamination of Clinical Photon Beams," P.L. Petti, M.S. Goodman, T.A. Gabriel, R. Mohan, Med. Phys., 10(1):18–24 (Jan.–Feb. 1983).

"Medium-energy Proton Accelerator for Therapy," B. Gottschalk, IEEE Transactions on Nuclear Science, NS-30(4):3063 (1983).

"Multi-dimensional Treatment Planning: I. Delineation of Anatomy," M. Goitein, M. Abrams, Int. J. Rad. Onc., Biol., Phys., 9:777–787 (1983).

"Multi-dimensional Treatment Planning: Ii. Beam's-eye View, Back Projection, And Projection through Ct Sections," M. Goitein, M. Abrams, D. Rowell, H. Pollari, J. Wiles, Int. J. Rad. Onc., Biol., Phys., 9:789–797 (1983).

"Planning Proton Therapy of the Eye," M. Goitein, T. Miller, Med. Phys., 10(3): 275–283 (May–June 1983).

"Problems of Inhomogeneities in Particle Beam Therapy," L. Verhey, M. Goitein, in *Pion and Heavy Ion Radiotherapy: Pre-clinical and Clinical Studies*, ed. L. Skarsgard, Elsevier Biomedical, New York-Amsterdam-Oxford, 159–168 (1983).

"Protons or Megavoltage X-rays as Boost Therapy For Patients Irradiated for Patients Irradiated for Localized Prostatic Carcinoma: An Early Phase I/Ii Comparison," J.R. Duttenhaver, W.U. Shipley, T. Perrone, L.J. Verhey, M. Goitein, J. Munzenrider, G.R. Prout, W.S. Kerr, Jr., E.C. Parkhurst, H.D. Suit, Cancer, 51:1599–1604 (1983).

"Proton Therapy at Harvard," J. Munzenrider, in *Pion and Heavy Ion Radiotherapy: Pre-clinical and Clinical Studies*, ed. L. Skarsgard, Elsevier Biomedical, New York-Amsterdam-Oxford, 363–372 (1983).

"Radiation Induced Soft Fails in Space Electronics," E.L. Peterson, IEEE Transactions on Nuclear Science, 1638–1641 (1983).

"Radiotherapy with 160 MeV Protons," B. Gottschalk, A.M. Koehler, R. Wilson, Proceedings of the Int. Symposium on Applications and Technology of Ionizing Radiations, 1, 181–203 (1983).

"Single Particle Effects in Microelectronics; A Problem in Microdosimetry," P. McNulty, Proc. of the Eighth Int. Symposium on Microdosimetry, ed. J. Booz, H. B. Ebert (Commission of the European Communities, Luxembourg), 921–933 (1983).

"Sources of Electron Contamination for the Clinac-35 25 MeV Photon Beam," P.L. Petti, M.S. Goodman, J. Sisterson, P. Biggs, T. Gabriel, R. Mohan, Med. Phys., 10(6):856–861 (Nov.–Dec. 1983).

"Determination of the Ca/P Molar Ratio Using Proton Activation Analysis," J. Sisterson, M. Grynpas, presented at the 25th Annual Meeting of the AAPM, New York (Aug. 1983). Abstract, Medical Physics, 10:735 (1983).

1984

"Bragg Peak Proton Beam Therapy for Arteriovenous Malformation of the Brain," Raymond N. Kjellberg, Kenneth R. Davis, Susan Lyons, William Rutler and Raymond D. Adams, Clinical Neurosurgery 31 (1984).

1985

"A measurement of W, the Energy Required to Create an Ion Pair, for 150 MeV Protons in Nitrogen and Argon," Paula L. Petti, Ph.D. Thesis (1985).

"Calculation of the uncertainty in the dose delivered during radiation therapy," M. Goitein, Medical Phyics, 12(5):608–612 (1985).

"Current Results of Proton Beam Irradiation of Uveal Melanomas," E. Gragoudas, J. Seddon, M. Goitein, L. Verhey, J. Munzenrider, M. Urie, H. Suit, P. Blitzer, A.M. Koehler, Ophthalmology, 92(2):284–291 (1985).

"Exact Calculation of Nonlinear Orbit Properties of a Synchrotron," B. Gottschalk, IEEE Transactions on Nuclear Science, NS-32(5) (Oct. 1985).

"Future Prospects in Planning Radiation Therapy," M. Goitein, Cancer, 55(9): 2234–2239 (1985).

"Hospital-based Accelerator for Proton Radiotherapy," B. Gottschalk, A.M. Koehler, M. Wagner, IEEE Transactions on Nuclear Science, NS-32:3305 (Oct. 1985).

"Methods for Calculating Seu Rates for Bipolar and Nmos Circuits," P.J. NcNulty, W. Abdel-Kader, J.M. Bisgrove, IEEE Trans. on Nuclear Science, NS-32(6), 4180–4184 (1985).

"Nonfunctioning Pituitary Adenoma," B. Kliman, R.N. Kjellberg, in Current Therapy in Endocrinology and Metabolism 1985–86, Krieger, Bardin, eds., Decker, Philadelphia, 21–26 (1985).

"Pathologic Examination of Ciliary Body Melanoma Treated with Proton Beam Irradiation," A. Ferry, C. Blair, E. Gragoudas, S. Volk, Archives of Ophtalmology, 103:1849–1853 (1985).

"Potential for Low-LET Charged Particle Radiation Therapy in Cancer," M. Goitein, H.D. Suit, E. Gragoudas, A.M. Koehler, R. Wilson, Radiation Research, 104:S-29 –S-309 (1985).

"Present and Future Proton Treatment Facilities at the Harvard Cyclotron Laboratory," J.M. Sisterson, K.N. Johnson, A.M. Koehler, B. Gottschalk, Nuclear Instruments and Methods in Physics Research B10/11, 1083–85 (1985).

"Progress in Low-LET Heavy Particle Therapy: Intracranial and Paracranial Tumors and Uveal Melanomas," M. Austin-Seymour, J. Munzenrider, M. Goitein, R. Gentry, E. Gragoudas, A.M. Koehler, P. McNulty, E. Osborne, D. Ryugo, J. Seddon, M. Urie, L. Verhey, H.D. Suit, Radiation Research, 104:S-219–S-226 (1985).

"Proton Beam Irradiation and Hyperthermia: Effects on Experimental Choroidal Melanoma," K.G. Riedel, P. Svitra, J. Seddon, D. Albert, E. Gragoudas, A.M. Koehler, J. Coleman, J. Torpey, F. Lizzi, J. Driller, Archives of Ophthalmology, 103:1862–1869 (Nov. 1985).

"Proton Therapy at Harvard," J. Munzenrider, M. Austin-Seymour, P. Blitzer, R. Gentry, M. Goitein, E. Gragoudas, K. Johnson, A.M. Koehler, P. McNulty, G. Moulton, E. Osborne, J. Seddon, H.D. Suit, M. Urie, L. Verhey, M. Wagner, Strahlentherapie, 161(12):756–763 (1985).

"Radiation Therapy for Malignant Intraocular Tumors," L. Brady, J. Shields, J. Augsburger, J. Day, A. Markoe, J. Castro, H. Suit. Strategies for treating possible tumor extension: Some theoretical considerations, M. Goitein, T. Schultheiss, Int. J. of Rad. Onc., Biol., Phys., 11:1519–1528 (1985).

"Proton Dose Distribution Influenced by Changes in Drift Space Between Patient and Bean-modifying Devices," M. Urie, M. Goitein, J.M. Sisterson, A.M. Koehler, J. Zoesman, presented at 27th Annual Meeting of the AAPM, Seattle,(Aug. 1985). Abstract, Medical Physics, 12:546 (1985).

"Vision Outcome in Parapapillary Uveal Melanomas After High-dose Precision Proton Beam Radiotherapy," J. Munzenrider, E. Gragoudas, J. Seddon, M. Goitein, M. Urie, L. Verhey, M. Austin-Seymour, J.-L. Habrand, E. Osborne, R. Gentry, S. Birnbaum, P. McNulty, K. Johnson, A. Koehler, K. Egan, L. Polivogianis, presented at meeting of Am. Soc. of Therapeutic Radiologists and Oncologists, Miami, FL (Oct. 1985).

1986

"A Measurement of W for 150 MeV Protons in Nitrogen and Argon," P. Petti, L. Verhey and Richard Wilson, Phys. Med. Biol., 31,10:1129–1138 (1986).

"As we Approach 3000: Proton Radiation Therapy at the Harvard Cyclotron Laboratory," J.M. Sisterson, K.N. Johnson, Am. Assoc. of Medical Dosimetrists, 1(1):21–30 (1986).

"Causes and Consequences of Inhomogeneous Dose Distributions in Radiation Therapy," M. Goitein, Int. J. of Rad. Onc., Biol, Phys., 12:701–704 (1986).

"Comparison of Soft Errors Induced by Heavy Ions and Protons," J.M. Bisgrove, J.E. Lynch, P.J. McNulty, W.G. Abdel-Kader, V. Kletnieks, W.A. Kolasinski, IEEE Trans. on Nuclear Science, NS-33(6):1571–1576 (1986).

"Degredation of the Bragg Peak Due to Inhomogeneities," M. Urie, M. Goitein, W.R. Holley, G.T.Y. Chen, Phys. Med. Biol., 31(1):1–15 (1986).

"Kombinierte Hyperthermie und Protonenbestrahlung: Ein Neuer Weg zur Behandlung des Malignen Aderhautmelanoms?, K. Riedel, P. Svitra, D. Albert, J. Seddon, E. Gragoudas, A.M. Koehler, D. Coleman, Fortschritte de Ophthalmologie, 83:483–488 (1986).

"Prognostic Factors for Metastasis Following Proton Beam Irradiation of Uveal Melanomas," E. Gragoudas, J. Seddon, K. Egan, L. Polivogianis, C.C. Hsieh, M. Goitein, L. Verhey, J. Munzenrider, M. Austin-Seymour, M. Urie, A.M. Koehler, Ophthalmology, 93(5):675–680 (May 1986).

"Proton Beam Penumbra: Effects of Separation Between Patient and Beam Modifying Devices," M. Urie, J.M. Sisterson, A.M. Koehler, M. Goitein, J. Zoesman, Medical Physics, 13(5):734–741 (Sept./Oct. 1986).

"Proton Beam Therapy Of Uveal Melanomas (Letter to the Editor)," E. Gragoudas, Archives of Ophthalmology, 104:349–351 (Mar. 1986).

"Radiation Therapy for Malignant Intraocular Tumors," L. Brady, J. Shields, J. Augsburger, J. Day, A. Markoe, J. Castro, H. Suit, in Diagnostic Imaging in Ophthalmology, ed. C. Gonzalez, M. Becker, J. Flanagan, Springer Verlag, New York, 343–357 (1986).

"Stereotactic Bragg Peak Proton Beam Radiosurgery for Cerebral Arteriovenous Malformations," R.N. Kjellberg, Annals of Clinical Research (Current Topics in Neurosurgery), 18(Supp. 47):17–19 (1986).

"Visual Outcome after Proton Beam Irradiation of Uveal Melanoma," J. Seddon, E. Gragoudas, L. Polivogianis, C.C. Hsieh, K. Egan, M. Goitein, L. Verhey, J. Munzenrider, M. Austin-Seymour, M. Urie, A.M. Koehler, Ophthalmology, 93(5):666–674 (May 1986).

"Monte Carlo Investigation of Compensating Bolus for Proton Beams: Influence of Air Gaps and Oblique Incidence," J.M. Sisterson, M. Urie, A.M. Koehler, M. Goitein, presented at 28th Annual Meeting of the AAPM, Lexington (Aug. 1986). Abstract, Medical Physics, 13:601 (1986).

"Preliminary Design Study for a Corkscrew Gantry," A.M. Koehler, presented at the Fifth PTCOG Meeting and Int. Workshop on Biomedical Accelerators, LBL-22962, UC-48, CONF-861271 (1986). (Also in Proceedings, 1987).

"Survival in Proton Irradiated Uveal Melanoma Patients: Implications for Prospective Randomized Trials," J. Munzenrider, E. Gragoudas, J. Seddon, P. McNulty, J. Sisterson, K. Johnson, M. Austin-Seymour, S. Birnbaum, R. Gnetry, A.M. Koehler, E. Osborne, D. Ruotolo, H.D. Suit, M. Urano, L. Verhey, presented at 28th Annual Meeting of ASTRO, Los Angeles, (Nov. 1986).

"25 years of Medical Physics at the Harvard Cyclotron Laboratory," J.M. Sisterson, K. Johnson, poster presented at the 28th Annual Meeting of the AAPM, Lexington (Aug. 1986). Abstract, Medical Physics, 13:577 (1986).

"The use of Protons in the Treatment of Malignancies of the Eye," J. Munzenrider, E. Gragoudas, J. Seddon, K. Johnson, presented at the 14th Int. Cancer Congress, Budapest (Aug. 1986).

1987

"Proton Beam Treatment Facility for Tumors of the Eye," A.M.Koehler, E. Gragoudas, B. Gottschalk, J. Munzenrider, J. Sisterson, M. Wagner,R. Wilson, presented at the Second International Meeting on Diagnosis andTreatment of Intraocular Tumors, Nyon, Switzerland (Nov. 1987).

"Clinical Results of Proton Beam Radiotherapy in Boston," J. Munzenrider,

M. Austin-Seymour, E. Gragoudas, J. Seddon, L. Verhey, M. Goitein, R. Gentry, M. Urie, D. Ruotolo, S. Birnbaum, K. Johnson, J. Sisterson, P. McNulty, H. Suit, A.M. Koehler, in Radiation Research, Proc. of 8th Int. Cong. of Radiation Research, Edinburgh, Taylor & Francis, London, New York, Philadelphia, 916–921 (1987).

"Design of a Hospital-based Accelerator for Proton Radiation Therapy: Scaling Rules," B. Gottschalk, Nucl. Instr. and Meth. in Physics Research B24/25:1092–1095 (1987).

"Effectiveness of CMOS Charge Reflection Barriers in Space Radiation Environments," P. McNulty, J. Lynch, W. Abdel-Kader, IEEE Trans. Nuclear Science, NS-34:1796–1799 (1987).

"Estimating the Dimensions of the SEU-sensitive Volume," W. Abdel-Kader, P. McNulty, S. El-Teleaty, J. Lynch, A. Khondker, IEEE Trans. Nuclear Science, NS-34:1300–1304 (1987).

"Long-term Results of Proton Beam Irradiated Uveal Melanomas," E. Gragoudas, J. Seddon, K. Egan, R. Glynn, J. Munzenrider, M. Austin-Seymour, M. Goitein, L. Verhey, M. Urie, A.M. Koehler, Ophthalmology, 94(4):349–353 (1987).

"Measurement of Cross-sections for Aluminum-26 and Sodium-24 Induced by Protons in Aluminum," R.J. Schneider, J.M. Sisterson, A.M. Koehler, Nucl. Inst. & Meth. in Physics Research, B29:271–274 (1987).

"Preliminary Results of Proton Beam Irradiation of Parapapillary Melanomas," E. Gragoudas, J. Seddon, K. Egan, M. Goitein, J. Munzenrider, L. Verhey, M. Austin-Seymour, M. Urie, A.M. Koehler, Proc. of Retina Workshop, Florence, Italy, 1986, Kugler & Ghedini, Amsterdam, Berkeley, Milan, 379–384 (1987).

"Present Results of Proton Beam Irradiation of Uveal Melanomas," E. Gragoudas, J. Seddon, K. Egan, M. Goitein, J. Munzenrider, L. Verhey, M. Austin-Seymour, M. Urie, A.M. Koehler, Proc. of the XXVth Int. Cong. of Ophthalmology, Rome, 1986, Kugler & Ghedini, Amsterdam, Berkeley, Milan, 911–916 (1987).

"Radiotherapie Mit Protonen (Proton Beam Therapy)," J. Munzenrider, M. Austin-Seymour, E. Gragoudas, J. Seddon, L. Verhey, M. Goitein, H. Suit, A.M. Koehler, presented at the Symposium on New Advances in the Radiation Therapy of Malignant Tumors, also in Wirkungssteigerung der Strahlentherapie Maligner Tumoren, ed. K. zum Winkel, Springer Verlag, Berlin, Heidelberg, New York, 58–68 (1987).

"Single-event, Enhanced Single-event and Dose-rate Effects with Pulsed Proton Beams," M. Xapsos, L. Massengill, W. Stapor, P. Shapiro, A. Campbell, S. Kerns, K. Fernald, A. Knudson, IEEE Transactions on Nuclear Science, NS-34(6):1070–1075 (Dec. 1987).

"Simulation of Low-earth-orbit Radiation Environments with a 5 to 120 MeV Proton Cyclotron Beam Using a Proton Beam Modulator," J. Suter, J. Cloeren, J. Norton, D. Kusnierkiewicz, A.M. Koehler, IEEE Transactions on Nuclear Science, NS-34(4):1070–1075 (Aug. 1987).

"Transposition of Target Information from the Magnetic Resonance and Ct-scan Images to the Conventional X-ray Stereotactic Space," A. De Salles, W. Asfora, M. Abe, R.N. Kjellberg, J. of Applied Neurophysiology, 50:23–32 (1987).

Uveal Melanomas Near the Optic Disc or Fovea: Visual Results after Proton Beam Irradiation, J. Seddon, E. Gragoudas, K. Egan, R. Glynn, J. Munzenrider, M. Austin-Seymour, M. Goitein, L. Verhey, M. Urie, A.M. Koehler, Ophthalmology, 94:354–361 (1987).

"Clinical Results of Fractionated Proton Beam Radiotherapy," J. Munzenrider, M. Austin-Seymour, E. Gragoudas, J. Seddon, J. Sisterson, presented at the 11th Clinical Cancer Research Conference: Biological Response Modifiers, Ontario (Sept. 1987). Published in Cancer in Ontario, 1987, The Ontario Cancer Research and Treatment Foundation, 27–34 (1987). Abstract in Int. J. of Radiation Onc., Biol., Phys., 13 Suppl.1:140 (1987).

"Measurement of Cross-sections for Aluminum-26 and Sodium-24 Induced by Protons in Aluminum," R.J. Schneider, J.M. Sisterson, A.M. Koehler, J. Klein, R. Middleton, poster presented at the 4th Int. Symposium on Accelerator Mass Spectrometry, Niagara-on-the-Lake, Ontario (Apr. 1987).

"Neurovisual Outcome Following Proton Radiation Therapy," J-L. Habrand, M. Austin-Seymour, S. Birnbaum, S. Wray, R. Carroll, J. Munzenrider, L. Verhey, M. Urie, M. Goitein, presented at the 29th Annual Meeting of ASTRO, Boston (Oct. 1987). Abstract in Int. J. of Radiation Onc., Biol., Phys., 13 Suppl. 1:141 (1987).

"Preliminary Study for a Corkscrew Gantry for Proton Beam Delivery," A.M. Koehler, J. Sisterson, B. Gottschalk, M.S.Z. Rabin, H. Enge, presented at the 29th Annual Meeting of the AAPM, Detroit (1987). Abstract, Medical Physics, 14:468 (1987).

"Proton Beam Treatment Facility for Tumors of the Eye," A.M. Koehler, E. Gragoudas, B. Gottschalk, J. Munzenrider, J. Sisterson, M. Wagner, R. Wilson, presented at Second International Meeting on Diagnosis and Treatment of Intraocular Tumors, Nyon, Switzerland (Nov. 1987).

"Proton Radiation Therapy: The Harvard Cyclotron Experience," J. Sisterson, K. Johnson, presented at the 29th Annual Meeting of the AAPM, Detroit (1987). Abstract, Medical Physics, 14:469 (1987).

"26 years of Proton Radiation Therapy at the Harvard Cyclotron Laboratory," J. Sisterson, K. Johnson, presented at the 29th Annual Meeting of ASTRO, Boston (Oct. 1987). Abstract in Int. J. of Radiation Onc., Biol., Phys., 13 Suppl. 1:174 (1987).

1988

"Air Gaps Between Bolus and Patient: Effects on the Compensation," J. Sisterson, M. Urie, A.M. Koehler, M. Goitein, HCL internal report (1988).

"Complications after Proton Beam Therapy for Uveal Malignant Melanoma, a Clinical and Histopathologic Study of Five Cases," M. Kincaid, R. Folberg, E. Torczynski, Z. Zakov, J. Shore, S. Liu, T. Planchard, T. Weingeist, Ophthalmology, 95:982–991 (1988).

"Conservative Treatment of Uveal Melanoma: Probability of Eye Retention After Proton Treatment," J. Munzenrider, E. Gragoudas, J. Seddon, J. Sisterson, P. McNulty, S. Birnbaum, K. Johnson, M. Austin-Seymour, J. Slater, M. Goitein, L. Verhey, M. Urie, D. Ruotolo, K. Egan, F. Osuna, Int. J. of Rad. Onc., Biol., Phys., 15:553–558 (1988).

"Endocrine Function Following High Dose Proton Therapy for Tumors of the Upper Clivus," J. Slater, M. Austin-Seymour, J. Munzenrider, S. Birnbaum, R. Carroll, A. Klibanski, P. Riskind, M. Urie, L. Verhey, M. Goitein, Int. J. of Rad. Onc., Biol., Phys., 15:607–611 (1988).

"Enucleation Versus Plaque Irradiation for Choroidal Melanoma," B. Straatsma, S. Fine, J. Earle, B. Hawkins, M. Diener-West, J. McLaughlin (COMS), Ophthalmology, 95:1000–1004 (1988).

"The Gamma Knife," R. Kjellberg, letter to the editor, J. of the Am. Medical Association, 260(17):2505 (1988).

"Heavy Particle Radiation Therapy," J. Munzenrider, M. Austin-Seymour, H. Suit, L. Verhey, in Clinical Radiation Therapy: Indications, techniques and results, ed. C.C. Wang, Wright Publishing, Littleton, MA, 411–432 (1988).

"Metastasis from Uveal Melanoma after Proton Beam Irradiation," E. Gragoudas, J. Seddon, K. Egan, R. Glynn, M. Goitein, J. Munzenrider, L. Verhey, M. Urie, A.M. Koehler, Ophthalmology, 95:992–999 (1988).

"Particle Radiation Therapy Research Plan," H. Suit, T. Griffin, J. Castro, L. Verhey, Am. J. Clinical Oncology, 11(3), 330–341 (1988).

"Potential for Improvement in Radiation Therapy," H. Suit, J. Becht, J. Leong, M. Stracher, W. Wood, L. Verhey, M. Goitein, Int. J. of Rad. Oncol., Biol., Phys., 14:777–786 (1988).

"The proton Accelerator in Neurosurgery," B. Swisher, S. Lyons, J. Vasapoli, in Nursing Yearbook 88, Springhouse Corp., Springhouse. PA, 80–83 (1988).

"Proton Beam Radiotherapy," L. Verhey, in Encyclopedia of Medical Devices and Instrumentation, ed. J.C. Webster, J. Wiley & Sons, New York, 4, 2354–2368 (1988).

"Proton Beam Therapy for Arteriovenous Malformations of the Brain," R. Kjellberg, in Operative Neurosurgical Techniques, Vol.2, Ch.81, ed. H. Schmidek, W. Sweet, Grune & Stratton, New York, 911–915 (1988).

"Radiation Therapy for Localized Prostate Carcinoma: Experience at the Massachusetts General Hospital (1973–1981)," W. Shipley, G. Prout, N. Coachman, P. McManus, E. Healey, A. Althausen, N. Heney, E. Parkhurst, H. Young, J. Shipley, S. Kaufman, NCI Monographs, No. 7, 67–73 (1988).

"Safety and Effectiveness of Magnetic Resonance Imaging of Choroidal Melanoma Patients with Episcleral Tantalum Rings after Proton Beam Irradiation," E. Smith, N. Kolodny, E. Gragoudas, L. Rubin, J. Seddon, D. D'Amico, Am. Journal of Ophthalmology, 105(6):695–696 (1988).

"The scope of the Problem of Primary Tumor Control," H. Suit, Cancer, 61(11): 2141-2147 (1988).

"Stereotactic Bragg Peak Proton Beam Therapy," R.N. Kjellberg, M. Abe, in Modern Stereotactic Neurosurgery, ed. L.D. Lunsford, Martinus Nijhoff Publishing, Boston/Dordrecht/Lancaster, 463–470 (1988).

"Uveal Melanoma: Conservative Treatment with Radiation Therapy," J. Munzenrider, E. Gragoudas, P. McNulty, J. Seddon, M. Urie, in Medical Radiology, Innovations in Radiation Oncology, ed. H. Withers, L. Peters, Springer Verlag, Berlin/Heidelberg, 41–50 (1988).

"Clinical Utilization of the Proton Beam: Past, Present and Future," A.M. Koehler, J.M. Sisterson, R. Wilson, M. Rabin, presented at meeting of the Philadelphia Chapter of the American Vacuum Society, Philadelphia (Feb. 1988).

"Compact Designs for Comprehensive Proton Beam Clinical Facilities," M. Rabin, B. Gottschalk, A.M. Koehler, J. Sisterson, presented at the 10th Conference on the Application of Accelerators in Research and Industry, Denton, TX (Nov. 1988). Abstract in Bulletin of the American Physical Society, 33(8):1763 (1988).

"Comparison of Variable and Fixed Modulation Proton Beam Dose Distributions in the Cranium," M. Urie, presented at the 30th Annual Meeting of the AAPM, World Congress on Medical Physics and Biomedical Engineering, San Antonio (Aug. 1988). Abstract in Physics in Medicine and Biology, 33:131 (1988).

"Conservative Treatment of Uveal Melanoma: Dose Distribution to Tumors with Local Recurrence after Proton Beam Therapy," J. Munzenrider, L. Verhey, E. Gragoudas, J. Seddon, P. McManus, S. Finn, J. Sisterson, K. Johnson, M. Urie, R. Gentry, S. Birnbaum, D. Ruotolo, C. Crowell, K. Egan, D. Lento, presented at 30th Annual Meeting of ASTRO, New Orleans (1988).

"Proton Radiation Therapy: Where will it be in the Year 2000?," J. Sisterson, K. Johnson, A.M. Koehler, J. Munzenrider, M. Rabin, presented at the 30th Annual Meeting of the AAPM, San Antonio (Aug. 1988). Abstract in Physics in Medicine and Biology, 33 Suppl. 1:131 (1988).

"A Test of Bragg-Gray Cavity Theory for Protons," L. Verhey, presented at the 30th Annual Meeting of the AAPM, World Congress on Medical Physics and Biomedical Engineering, San Antonio (Aug. 1988). Abstract in Physics in Medicine and Biology, 33:73 (1988).

1989

"Clinical Use of Protons and Ion Beams from a World-wide Perspective," J. Sisterson, Nuclear Instruments and Methods in Physics Research, B40/41:1350–1353 (1989).

"Compact Designs for Comprehensive Proton Beam Clinical Facilities," M. Rabin, B. Gottschalk, A.M. Koehler, J. Sisterson, L. Verhey, Nuclear Instruments and Methods in Physics Research, B40/41:1335–1339 (1989).

"Conservative Treatment of Uveal Melanoma: Local Recurrence after Proton Beam Therapy," J. Munzenrider, L. Verhey, E. Gragoudas, J. Seddon, M. Urie, R. Gentry, S. Birnbaum, D. Ruotolo, C. Crowell, P. McManus, S. Finn, J. Sisterson, K. Johnson, K. Egan, D. Lento, P. Bassin, Int. J. Rad. Onc., Biol. Phys., 17:493–498 (1989).

"Distal Penetration of Proton Beams: The Effect of Air Gaps Between Compensating Bolus and Patient," J. Sisterson, M. Urie, A.M. Koehler, M. Goitein, Phys. Med. Biol., 34:1309–1315 (1989).

"Evaluation of Tumor Regression and other Prognostic Factors for Early and Late Metastasis after Proton Irradiation of Uveal Melanoma," R. Glynn, J. Seddon, E. Gragoudas, K. Egan, L. Hart, Ophthalmology, 96(10):1566–1573 (1989).

"Fractionated Proton Radiation Therapy of Chordoma and Low Grade Chondrosarcoma of the Base of Skull," M. Austin-Seymour, J. Munzenrider, M. Goitein, L. Verhey, M. Urie, R. Gentry, S. Birnbaum, D. Ruotolo, P. McManus, S. Skates, R. Ojemann, A. Rosenberg, A. Schiller, A.M. Koehler, H. Suit, Journal of Neurosurgery, 70:13–17 (1989).

"The Influence of the Size of the Grid Used for Dose Calculation on the Accuracy of Dose Estimation," A. Niemierko, M. Goitein, Medical Physics, 16(2):239–247 (1989).

"Neurovisual Outcome Following Proton Radiation Therapy," J.-L. Habrand, M. Austin-Seymour, S. Birnbaum, S. Wray, R. Carroll, J. Munzenrider, L. Verhey, M. Urie, M. Goitein, Int. J. Radiation Oncology, Biology, Physics, 16:1601–1606 (1989).

"Proton Beam Therapy – Cost vs. Benefit," F. Antoine, Journal of the National Cancer Institute, 81(8):559–562 (1989).

"Proton Therapy for Carcinoma of the Nasopharynx: A Study in Comparative Treatment Planning," A.Brown, M. Urie, R. Chisin, H. Suit, Int. J. Radiation Oncology, Biology, Physics, 16:1607–1614 (1989).

"The Risk of Enucleation after Proton Beam Irradiation of Uveal Melanoma," K Egan, E. Gragoudas, J. Seddon, R. Glynn, J. Munzenrider, M. Goitein, L. Verhey. M. Urie, A.M. Koehler, Ophthalmology, 96(9):1377–1383 (1989).

"Variable vs. Fixed Modulation of Proton Beam for Treatments in the Cranium," M. Urie, M. Goitein, Medical Physics, 16(4):593–601 (1989).

1990

"Fractionated Proton Radiation Therapy of Cranial and Intracranial Tumors,"Austin-Seymour, M., J. Munzenrider, et al. (1990), Am. J. of Clinical Oncology 13(4):327–330.

"Considerations in Fractionated Proton Radiation Therapy: Clinical Potential and Results," Austin-Seymour, M., M. Urie, et al. (1990), Radiotherapy and Oncology 17:29–35.

"Production of Ca^{41} and K, Sc and V Short-lived Isotopes by the Irradiation of Ti with 35 to 150 MeV Protons: Applications to Solar Cosmic Ray Studies," Fink, D., J. Sisterson, et al. (1990), Nucl. Instr. and Meth. in Phys. Res. B52:601–607.

"Evaluation of Treatment Planning for Particle Beam Radiotherapy," Goitein, M., M. Urie, et al. (1990), Int J Rad Onc Biol Phys 21.

"Resetting a Current Integrator with the Supply Lines. [Letter to Editor]," Gottschalk, B. (1990), Nuclear Instruments and Methods in Physics Research A297:534–535.

"Relative Survival Rates after Alternative Therapies for Uveal Melanoma," Seddon, J., E. Gragoudas, et al. (1990), Ophthalmology 97(6):769–777.

"Overview of Proton Beam Applications in Therapy," Sisterson, J. (1990), Nucl. Inst. and Meth. in Phys. Res. B45:718–723.

"Increased Efficacy of Radiation Therapy by Use of Proton Beam," Suit, H., M. Goitein, et al. (1990), Strahlentherapie und Onkologie 166:40–44.

"Random Sampling for Evaluating Treatment Plans," Niemierko, A. and M. Goitein (1990), Med. Phys. 17(5):753–762.

"Evaluation of New Radiation Oncology Technology," Smith, A. (1990), Int J Rad Oncol Biol Phys 18:701–703.

"The role of Radiation Therapy. Tumors of the Spine," Suit, H. and M. Austin-Seymour (1990), N. Sundaresan, H. Schmidek, A. Schiller and D. Rosenthal. Philadelphia, PA, WB Saunders: 86–91.

"Overview of Proton Beam Applications in Therapy," J. Sisterson, Phys. Res. B45:718–723 (1990).

"Considerations in Fractionated Proton Radiation Therapy: Clinical Potential and Results," M. Austin-Seymour, M. Urie, C. Willett, M. Goitein, L. Verhey, R. Gentry, P. McNulty, A. Koehler, H. Suit, Radiotherapy and Oncology, 17:29–35 (1990).

"Fractionated Proton Radiation Therapy of Cranial and Intracranial Tumors," M. Austin-Seymour, J. Munzenrider, R. Linggood, M. Goitein, L. Verhey, M. Urie, R. Gentry, S. Birnbaum, D. Ruotolo, C. Crowell, P. McManus, S. Skates, A.M. Koehler, H. Suit, Am. J. of Clinical Oncology, 13(4):327–330 (1990).

"Production of CA^{41} and K, Sc and V Short-lived Isotopes by the Irradiation of Ti with 35 to 150 MeV Protons: Applications to Solar Cosmic Ray Studies," D. Fink, J. Sisterson, S. Vogt, G. Herzog, J. Klein, R. Middleton, A. Koehler, Nuclear Instruments and Methods in Physics Research, B52:601–607 (1990).

"Average SCR Flux During Past 105 Years: Inference from Ca^{41} in Lunar Rock 74275," J. Klein, D. Fink, R. Middleton, S. Vogt, G. Herzog, R. Reedy, J. Sisterson, A.M. Koehler, A. Meglis, presented at Lunar and Planetary Science Conference XXI (March 1990).

1991

"3-D Comparative Study of Proton vs. X-ray Radiation Therapy for Rectal Cancer," Tatsuzaki, H., M. Urie, et al. (1991), Int. J. of Rad. Onc., Biol., Phys. 22:369–374.

"Comparative Treatment Planning: Proton vs. X-ray Beams against Glioblastoma Multiforme," Tatsuzaki, H., M. Urie, et al. (1991), Int. J. of Rad. Onc., Biol., Phys. 22:265–273.

"Charge Collection Measurements and Theoretical Calculations for Partially Depleted Silicon Devices," Abdel-Kader, W., S. El-Teleaty, et al. (1991), Nucl. Instr. and Meth. in Phys. Res. B 56–57:1246–1250.

"Biopsy after External Beam Radiation Therapy for Adenocarcinoma of the Prostate: Correlation with Original Histological Grade and Current Prostate Specific Antigen Levels," Dugan, T., W. Shipley, et al. (1991), J. of Urology 146:1313–1316.

"Charge Collection in Partially Depleted GaAs Test Structures Induced by Alphas, Heavy Ions, and Protons," El-Teleaty, S., W. Abdel-Kader, et al. (1991), J. Appl. Phys. 69(1):475–480.

"Multiple Scattering of Protons in Compounds and Thick Targets: Tests of Molière Theory," Gottschalk, B., A. M. Koehler, et al. (1991), 4th Workshop on Heavy Charged Particles in Biology and Medicine and 15th Meeting of PTCOG, Darmstadt, Germany, Proceedings.

"The Case for Passive Beam Spreading," Gottschalk, B., A. M. Koehler, et al. (1991), Proton Therapy Workshop, Paul Scherrer Institute, Villigen, Switzerland, Proceedings.

"Survival of Patients with Metastases from Uveal Melanoma," Gragoudas, E., K. Egan, et al. (1991), Ophthalmology 98(3):383–389.

"Proton Irradiation of Uveal Melanomas: The First 1000 Patients," Gragoudas, E., J. Seddon, et al. (1991), Second International Symposium on Diagnosis and Treatment of Intraocular Tumors, Geneva, Nov. 1987, Amsterdam, New York, Kugler.

"Faraday Cup Dosimetry in Proton Beam Therapy [Abstract]," Mayo, C., J. M. Sisterson, et al. (1991), 33rd Annual Meeting of the AAPM, San Francisco, CA, Medical Physics.

"Modeling Charge Collection and Single Event Upsets in Microelectronics," McNulty, P., W. Abdel-Kader, et al. (1991), Nucl. Instr. and Meth. in Phys. Res. B 61:52–60.

"Potential Improvement of Three Dimension Treatment Planning and Proton Therapy in the Outcome of Maxillary Sinus Cancer," Miralbell, R., C. Crowell, et al. (1991), Int. J. of Rad. Onc., Biol., Phys. 22:305–310.

"Proton Beam Therapy: Reliability of the Synchrocyclotron at the Harvard Cyclotron Laboratory," Sisterson, J., E. Cascio, et al. (1991), Phys. Med. Biol. 36(2):285–290.

"Cross section Measurements at the Harvard Cyclotron Laboratory," Sisterson, J. and A. M. Koehler (1991), Advisory Group Meeting, IAEA, Vienna, Austria, INDC (NDS)–245.

"Cross Sections for Production of Carbon-14 from Oxygen and Silicon: Implications for Cosmogenic Production Rates," Sisterson, J. M., A. Jull, et al. (1991), 54th Meteoritical Society Meeting, Monterey, CA, Meteoritics.

"Proton Beams in Clinical Radiation Therapy," Suit, H. (1991), J. of the Japanese Soc. of Therapeutic Radiol. Oncol. 3:191–198. Not Available.

"Proton Beam Irradiation for Treatment of Experimental Human Retinoblastoma," Svitra, P., D. Budenz, et al. (1991), Eur. J. of Ophthalmology 1(2):57–62.

"Importance of Precise Positioning for Proton Beam Therapy in the Base of Skull and Cervical Spine," Tatsuzaki, H. and M. Urie (1991), Int. J. of Rad. Onc., Biol., Phys. 21:757–765.

"Fitting of Normal Tissue Tolerance Data to an Analytic Function," Burman, C., G. Kutcher, et al. (1991), Int J Radiat Oncol Biol Phys 21:123–135.

"BPM Data Acquisition System for the Bates Pulse Stretcher Ring. [Abstract]," Calvo, O., T. Russ, et al. (1991), Particle Accelerator Conference.

"Three-dimensional Photon Treatment Planning in Carcinoma of the Larynx," Coia, L., J. Gavin, et al. (1991), Int J Radiat Oncol Biol Phys 21:183–192.

"Dose-volume Histograms," Drzymala, R., R. Mohan, et al. (1991), Int J Radiat Oncol Biol Phys 21:71–78.

"Tolerance of Normal Tissue to Therapeutic Irradiation," Emami, B., J. Lyman, et al. (1991), Int J Radiat Oncol Biol Phys 21:109–122.

"Energy Compression System Design for the MIT-Bates Accelerator Center. [Abstract]," Flanz, J., P. Demos, et al. (1991), Particle Accelerator Conference.

"Plans for a Proton Medical Facility at Massachusetts General Hospital," Gall, K., M. Goitein, et al. (1991), NIRS International Workshop on Heavy Charged Particle Therapy and Related Subjects, Chiba, Japan.

"Radiosensitization with Carotid Arterial Infusion of Bromodeoxyuridine

(BUdR) +/– 5 Fluorouracil (FU[5]) Biomodulation with Focal External Beam Radiation (FEBT) for Malignant Gliomas. [Abstract]," Greenberg, H. S., W. F. Chandler, et al. (1991), International Symposium on Advances in Neuro-Oncology, Samremo, Italy, Kluwer Academic Publishers, B.V., The Netherlands.

"Instability Calculations for the MIT-Bates SHR [Abstract]," Jacobs, K., P. Demos, et al. (1991), Particle Accelerator Conference.

"Histogram Reduction Method for Calculating Complication Probabilities for Three-dimensional Treatment Planning Evaluations," Kutcher, G. J., C. Burman, et al. (1991), Int J Radiat Oncol Biol Phys 21:137–146.

"Malignant Astrocytomas: Focal Tumor Recurrence after Focal External Beam Radiation Therapy (FBRT)," Liang, B. C., A. F. Thornton, et al. (1991), J. Neurosurgery 74(4):559–563.

"Three-dimensional Treatment Planning for Para-aortic Node Irradiation in Patients with Cervical Cancer," Munzenrider, J., K. Doppke, et al. (1991), Int J Radiat Oncol Biol Phys 21:229–242.

"Numerical Scoring of Treatment Plans," Munzenrider, J. E., G. J. Kutcher, et al. (1991), Int J Radiat Oncol Biol Phys 21:147–163.

"Calculation of Normal Tissue Complication Probability and Dose-volume Histogram Reduction Schemes for Tissues with a Critical Element Architecture," Niemierko, A. and M. Goitein (1991), Radiother & Oncol 20:166–176.

"Three-dimensional Photon Treatment Planning," Smith, A. R. and J. A. Purdy, Eds. (1991), Report of the Collaborative Working Group on the Evaluation of Treatment Planning for External Photon Beam Radiotherapy. [Editor's Note]. Int J Radiat Oncol Biol Phys.

"Three-dimensional Photon Treatment Planning of the Intact Breast," Solin, L. J., J. C. H. Chu, et al. (1991), Int J Radiat Oncol Biol Phys 21:193–203.

"Three-Dimensional Treatment Planning of Astrocytomas: A Dosimetric Study of Cerebral Irradiation," Thornton, A. F., T. J. Hegarty, et al. (1991), Int J Radiat Oncol Biol Phys 20:1309–1315.

"Three-Dimensional Motion Analysis of an Improved Head Immobilization System for Simulation, CT, MRI, and PET Imaging," Thornton, A. F., R. K. TenHaken, et al. (1991), Radiotherapy and Oncology 20:224–28.

"A Head Immobilization System for Radiation Simulation, CT, MRI, and PET Imaging," Thornton, A. F., R. K. TenHaken, et al. (1991), Medical Dosimetry 16:51–56.

"Protracted Lhermitte's Sign Following Head and Neck Irradiation," Thornton, A. F., S. Zimberg, et al. (1991), Arch Otolaryngol 117:1300–1303.

"The Role of Uncertainty Analysis in Treatment Planning," Urie, M., M. Goitein, et al. (1991), Int J Radiat Oncol Biol Phys 21:91–107.

"Design of MRI Scan Protocols for Use in 3–D, CT-Based Treatment Planning," Yanke, B. R., R. K. TenHaken, et al. (1991), Medical Dosimetry 16(4):205–211.

"The Importance of Optimal Treatment Planning in Radiation Therapy," Suit, H. and W. DuBois (1991), Int J Radiat Oncol Biol Phys (F. Ellis dedicatory issue) 21:1471–1478.

"Proton Beam Therapy: Reliability of the Synchrocyclotron at the Harvard Cyclotron Laboratory" J. Sisterson, E. Cascio, A.M. Koehler, K. Johnson, Phys. Med. Biol., 36(2):285–290 (1991).

"Determination of Solar-proton Fluxes Using Carbon-14 in Lunar Rocks," J. Sisterson, H. Román, J. Vogel, J. Southon, R. Reedy, presented at 22nd Lunar and Planetary Science Conference, Houston, TX (March 1991).

1992

"Modeling of Normal Tissue Response to Radiation: The Critical Volume Model," Niemierko, A. and M. Goitein (1992), Int J Radiat Oncol Biol Phys 25(1):135–145.

"Evaluation of Multileaf Collimator Design for a Photon Beam." Galvin, J., A. Smith, et al. (1992), Int J Radiat Oncol Biol Phys 23:789–801.

"The Comparison of Treatment Plans," Goitein, M. (1992), Seminars in Radiation Oncology 2(4):246–256.

"Optimization of 3D Radiation Therapy with Both Physical and Biological End Points and Constraints," Niemierko, A., M. Urie, et al. (1992), Int J Radiat Oncol Biol Phys 23:99–108.

"Functional MRI and PET FDG Studies: Correlations with Glial Tumor Prognosis," Pardo, F., H. Aronen, et al. (1992), IJROBP, proc. suppl.

"Improved methods for dEtermination of Variability in Patient Positioning for Radiation Therapy Using Simulation and Serial Portal Film Measurements," Rosenthal, S., J. Galvin, et al. (1992), Int J Rad Oncol Biol Phys 23:621–625.

"Spinal Cord Astrocytomas – Results of Therapy," Sandler, H., S. Papadopoulos, et al. (1992), Neurosurgery 30(4):490–493.

"Clinical Implications of Heterogeneity of Tumor Response oo Radiation Therapy," Suit, H., S. Skates, et al. (1992), Radiother & Oncol 25:251–260.

"Proton Beams in Clinical Radiation Therapy," Principles & Practice of Oncology – PPO Update," Suit, H., M. Urie, et al. (1992). V. DeVita, S. Hellman and S. Rosenberg. 6:1–15.

"Clinical Gains to be Realized from Proton Beams in Radiation Therapy," Suit, H. and M. Urie (1992), JNCI 84:155–164.

"Proton Beams in Radiation Therapy," Suit, H. and M. Urie (1992), JNCI 84(3):155–163.

"Local Control and Patient Survival," Suit, H. (1992), Int J Radiat Oncol Biol Phys 23:653–660.

"Potential for Proton Beams in Clinical Radiation Oncology," Suit, H. (1992),[ICRR presentation 1991]. Radiation Research, A Twentieth-Century Perspective. Vol. II: Congress Proceedings. W. Dewey, M. Edington, R. Fry, E. Hall and G. Whitmore. San Diego, CA, Academic Press, Inc.: 3–13.

"A Quantitative Assessment of the Addition of MRI to CT-Based 3-D Treatment Planning of Brain Tumors," TenHaken, R., A. Thornton, et al. (1992), Radiotherapy and Oncology 25:121–133.

"Current and Potential Clinical Indications of Proton Irradiation," Thornton, A. and H. Suit (1992), XIII International Congress on Cyclotrons, Vancouver, Canada, World Scientific.

"Recent Technical Advances in the Irradiation of Head and Neck Neoplasia," Thornton, A., K. Gall, et al. (1992), 3rd International Conference on Head and Neck Cancer, San Francisco, Elsevier Scientific.

"The Clinical Utility of Magnetic Resonance Imaging in 3-Dimensional Treatment Planning of Brain Neoplasms," Thornton, A., M. Sandler, et al. (1992), Int J Rad Oncol Biol Phys 24(4):767–776.

"A Dose Response Analysis of Injury to Cranial Nerves and/or Nuclei Following Proton Beam Radiation Therapy," Urie, M., B. Fullerton, et al. (1992), Int. J. of Rad. Onc., Biol., Phys 23:27–39.

"Radiation-Induced Optic Neuropathy: Correlation of MRI and Radiation Dosimetry," Young, W., A. Thornton, et al. (1992), Radiology 185:904–907.

"Histopathology of Proton Bean-irradiated vs Enucleated Uveal Melanomas," Saornil, M., K. Egan, et al. (1992), Arch. Ophthal. 110:1112–1118.

"A Cranial Immobilization and Repositioning System for Combined Proton and Photon Irradiation Of Paranasal Sinus Tumors [Abstract]." Rosenthal, S. and A. Thornton (1992), 16th Meeting of PTCOG, Vancouver, BC, Canada, Abstracts of the PTCOG XVI Meeting.

"CRRES Spectrometer for Electrons And Protons," Nightingale, R., R. Vondrak, et al. (1992), Journal of Spacecraft and Rockets 29 (4):614–617.

"The role of Radiotherapy in the Treatment of Subtotally Resected Benign Meningiomas," Miralbell, R., R. Linggood, et al. (1992), J. of Neuro-Oncology 13:157–164.

"Temperature Coefficient of Open Thimble Chambers," Mayo, C. and B. Gottschalk (1992), Phys. Med. Biol. 37(1):289–291.

"Stereotactic Alignment for Bragg Peak Radiosurgery," Butler, W., C. Ogilvy, et al. (1992), Radiosurgery: Baseline and Trends. L. Steiner. New York, Raven Press: 85–91.

"3D Treatment Planning for Heavy Charged Particles," Goitein, M. (1992), Fourth Workshop on Heavy Charged Particles in Biology and Medicine and XV PTCOG Meeting, Darmstadt, Radiat Environ Biophys.

"Uses of 3D Planning in Addition to Creating a Good Treatment: Ongoing Studies at MGH/HCL," Urie, M. M. (1992), Radiat Environ Biophys 31:247–250.

"Random Search Algorithm (Ronsc) for Optimization of Radiation Therapy with Both Physical and Biological End Points and Constraints," Niemierko, A. (1992), Int. J. Radiat. Oncol. Biol. Phys. 23(1):89–98.

1993

"Probable Causes of Recurrence in Patients with Chordoma and Chondrosarcoma at the Base of Skull and Cervical Spine," Austin, J., M. Urie, et al. (1993), Int J Radiat Oncol Biol Phys 25:439–444.

"Late Rectal Bleeding Following Combined X-ray and Proton High Dose Irradiation for Patients with Stages T3-t4 Prostate Carcinoma," Benk, V., J. Adams, et al. (1993), Int J Radiat Oncol Biol Phys 26:551–557.

"Prognostic Factors in Intracranial Neoplasms," Efird, J., D. Hsu, et al. (1993), Social Statistics Section, American Statistical Association, Alexandria, VA.

"State of the Art? New Proton Medical Facilities for the Massachusetts General Hospital and the University of California Davis Medical Center," Gall, K., L. Verhey, et al. (1993), Nucl. Instr. Meth. B79:881–84.

"Computer-assisted Positioning of Radiotherapy Patients Using Implanted Radio-opaque Fiducials," Gall, K., L. Verhey, et al. (1993), Med Phys 20(4):1153–1159.

"Multiple Coulomb Scattering of 160 MeV Protons," Gottschalk, B., A. Koehler, et al. (1993), Nuclear Instruments and Methods in Physics Research B74:467–490.

"Automated Determination of Patient Setup Errors in Radiation Therapy Using Spherical Radio-opaque Markers," Lam, K., R. TenHaken, et al. (1993), Med Phys 20(4):1145–1152.

"Alternating and Concurrent Chemotherapy and Radiotherapy for Unresectable Head and Neck Carcinoma," McLaughlin, P., S. Urba, et al. (1993), Radiation Oncology Investigations 1:111–116.

"Potential Improvement of Three-dimensional Treatment Planning and Proton Beams in Fractionated Radiotherapy of Large Cerebral Artereovenous Malformations," Mirabell, R. and M. Urie (1993), Int J Radiat Oncol Biol Phys 25:353–358.

"Chordoma and Chondrosarcoma of Skull Base: Treatment with Fractionated X-ray and Proton Radiotherapy," Munzenrider, J., N. Liebsch, et al. (1993), J. Johnson and M. Didolkar, Head and Neck Cancer, Elsevier Science Publishers B.V. 3:649–654.

"Implementation of a Model for Estimating Tumor Control Probability for an Inhomogeneously Irradiated Tumor," Niemierko, A. and M. Goitein (1993), Radiotherapy and Oncology 29(2):140–147.

"Intrinsic Radiation Sensitivity may not be the Major Determinant of the Poor Clinical Outcome of Glioblastoma Multiforme," Taghian, A., J. Ramsay, et al. (1993), Int J Radiat Oncol Biol Phys 25:243–249.

"Neuro-ophthalmologic Findings in Chordoma and Chondrosarcoma of the Skull Base," Volpe, N., N. Liebsch, et al. (1993), American Journal of Ophthalmology 115:97–104.

"Initiation of Multileaf Collimator Conformal Radiation Therapy," Powlis, W., A. Smith, et al. (1993), Int J Rad Oncol Biol Phys 25:171–179.

"Making Weapons to Fight Cancer," Koepfer, C. (1993), Modern Machine Shop 65(8):80–90.

"An Evaluation of the Influence of Reproductive Factors on the Risk of Metastases from Uveal Melanoma," Egan, K., S. Walsh, et al. (1993), Ophthalmology 100(8):1160–1166.

"Proton Therapy in 1993," Sisterson, J. M. (1993), XVIII PTCOG Meeting, Orsay and Nice, France, Proceedings.

"Uveal melanomas: Trends in patient referral for Proton Beam Therapy at HCL," Sisterson, J. M., K. N. Johnson, et al. (1993), XVIII PTCOG Meeting, Orsay and Nice, France, Proceedings.

"On the Sampling Techniques for the Evaluation of Treatment Plans, [Letter to the Editor]," Niemierko, A. and M. Goitein (1993), Med Phys 20(5):1377–1380.

1994

"Chondrosarcomas of the Skull Base," Brown, E., E. Hug, et al. (1994), Clinics of North America 4:529–544.

"Charged Particle Therapy for Base of Skull Tumors: Past Accomplishments and Future Challenges. [Editorial]," Hug, E. and J. Munzenrider (1994), Int J Radiat Oncol Biol Phys 29(4):911–912.

"Soft Tissue Sarcomas of the Head and Neck: Results of Combined Proton and Photon Radiation Therapy Using 3d Treatment Planning," Hug, E., P. Hanssens, et al. (1994), Int J Radiat Oncol Biol Phys 30(S1):222.

"Proton Therapy with Harvard Cyclotron," Munzenrider, J. (1994). 1st International Symposium on Hadrontherapy, Como, Italy, Hadrontherapy in Oncology, Elsevier Science.

"Charged Particles. Radiation Oncology, Technology, and Biology" Munzenrider, J. and C. Crowell (1994). P. Mauch and J. Loeffler, W.B. Saunders Company, pp. 34–55.

"Dose-volume Distributions (DVD's): A New Approach to Dose-volume Histograms in Three-dimensional Treatment Planning," Niemierko, A. and M. Goitein (1994), Med Phys 21(1):3–11.

"Dose-volume Effects in the Spinal Cord. [Letter to the Editor]." Niemierko, A. and M. Goitein (1994), Radiotherapy and Oncology 31(3):265–267.

"A New Approach to Optimization of Beam Intensity Profiles." Niemierko, A. and M. Goitein (1994), 36th AAPM Annual Meeting, Anaheim, CA.

"Base of Skull Chordoma: A Correlative Study of Histological and Clinical Features in 62 Cases," O'Connell, J., L. Renard, et al. (1994), Cancer 74(8):2261–2267.

"Is Tumor Cell Radiation Resistance Correlated with Metastatic Ability," Suit, H., A. Allam, et al. (1994), Cancer Research 54:1736–41.

"Role of Radiation in the Management of Adult Patients with Sarcoma of Soft Tissue," Suit, H. and I. Spiro (1994), Seminars in Surg Oncol 10:347–356.

"The Role of Radiation in Patients with Soft Tissue Sarcomas." Suit, H. and I. Spiro (1994), Cancer Control – Journal of the Moffitt Cancer Ctr 1(6):592–598.

"Principles of Radiation Oncology," Suit, H. and M. Urano (1994), in *Oxford Textbook of Surgery*. P. Morris and R. Malt. Oxford, Oxford Univ. Press: 2627–2634.

"World Experience in Proton/Ion Therapy in 1994," Sisterson, J, (1994). Proceedings of NIRS International Seminar on the Application of Heavy Ion Accelerator to Radiation Therapy of Cancer. T. Kanai and E. Takada. Chibashi, Japan, National Institute of Radiological Sciences: 83–85.

"Proton Production Cross Sections for C^{14} from Silicon and Oxygen: Implications for Cosmic-ray Studies," Sisterson, J., A. Jull, et al. (1994), Nucl. Instr. and Meth. in Phys. Res. B. 92:510-512.

"Biological Bases of 3D Treatment Plan Optimization." Niemierko, A. (1994), EUTECH 94 Conference, Genova, Italy, Proceedings.

"Estimates of Neutron Dose to Patients Receiving Proton Beam Treatment at Harvard Cyclotron Laboratory." [Abstract]. Koehler, A. M., X. Yan, et al. (1994), XX Meeting of PTCOG, Chester, England, Proceedings.

"Functional Cerebral Imaging in the Evaluation and Radiotherapeutic Treatment Planning of Patients with Malignant Glioma." Pardo, F. S., H. J. Aronen, et al. (1994), Int. J. Radiat. Oncol. Biol. Phys. 30:663–9.

"Familial Uveal Melanoma," Young, L., K. M. Egan, et al. (1994), Am. J. Ophthalmol. 117:516–20.

1995

"Base of Skull and Cervical Spine Chordomas in Children Treated by High-dose Irradiation," Benk, V., N. Liebsch, et al. (1995). Int J Radiat Oncol Biol Phys 31:577–581.

"Radiation Therapy for Chordomas of the Base of Skull and Cervical Spine:

Patterns of Failure and Outcome After Relapse." Fagundes, M., E. Hug, et al. (1995). Int J Radiat Oncol Biol Phys 33(3):579–584.

"Lens Changes after Proton Irradiation for Uveal Melanoma," Gragoudas, E., K. Egan, et al. (1995), Am J Ophthalmol 119:157–164.

"Locally Challenging Osteo- and Chondrogenic Tumors of the Axial Skeleton: Results of Combined Proton and Photon Radiation Therapy Using 3-d Treatment Planning," Hug, E., M. Fitzek, et al. (1995), Int J Radiat Oncol Biol Phys 31(3):467–476.

"Combined Surgery and Radiotherapy for Conservative Management of Soft Tissue Sarcomas," Hug, E., I. Spiro, et al. (1995), in *Soft Tissue Sarcomas in Adults, Recent Results in Cancer Research*. M. Bamberg + Berlin, Springer Verlag. 138:47–56.

"Skull Base Chordomas: Treatment Outcome and Prognostic Factors in Adult Patients Following Conformal Treatment with 3d Planning and High Dose Fractionated Combined Proton and Photon Radiation Therapy," Munzenrider, J., E. Hug, et al. (1995), J Radiat Oncol Biol Phys 32(S1):209.

"Radiation Dose Response of Human Tumors," Okunieff, P., D. Morgan, et al. (1995), J Radiat Oncol Biol Phys 32(4):1227–1237.

"A precision Cranial Immobilization System for Conformal Sterotactic Fractionated Radiation Therapy," Rosenthal, S., K. Gall, et al. (1995), Int J Radiat Oncol Biol Phys 33(5):1239–1245.

"Comparison of Proton and X-ray Conformal Dose Distributions for Radiosurgery Applications," Serago, C., A. Thornton, et al. (1995), Med Phys 22(12):2111–2116.

"Advanced Prostate Cancer: The Results of a Randomized Comparative Trial of High Dose Irradiation Boosting with Conformal Protons Compared with Conventional Dose Irradiation Using Photons Alone," Shipley, W., L. Verhey, et al. (1995), Int J Radiat Oncol Biol Phys 32(1):3–12.

"Proton Radiation Therapy: A Summary of the World Wide Experience," Sisterson, J. (1995), 13th Int. Conf. on Applications of Accelerators in Research and Industry, Nucl Instr Meth.

"Combined Surgery and Radiation Therapy for Limb Preservation in Soft Tissue Sarcoma of the Extremity," Spiro, I., A. Rosenberg, et al. (1995), Cancer Investigation 13(1):86–95.

"Soft tissue sarcomas. Treatment of Cancer," Suit, H., A. Rosenberg, et al. (1995), P. Price and K. Sikora. London, England, Chapman & Hall Medical:795–823.

"Sarcoma of Soft Tissues: Radiation Sensitivity, Treatment Field Margins, Pathological Margins, and Dose," Suit, H. (1995), Int J Rad Oncol Biol Phys 39:969–76.

"Radiation as a Therapeutic Modality in Sarcomas of the Soft Tissue," Suit, H. and I. Spiro (1995), Hematology/Oncology Clinics of North America.

S. Patel and R. Benjamin. Philadelphia, WB Saunders Co. 9:733–746.

"Sarcoma of the Soft Tissues," Suit, H., C. v. Groeningen, et al. (1995), in *Oxford Textbook of Oncology*. M. Peckham, H. Pinedo and U. Veronesi. Oxford, Oxford University Press: 1917–1939. Not Available.

"Tumor Oxygenation and Radiosensitivity. Blood Substitutes: Physiological Basis of Efficacy," Suit, H. (1995), Winslow. Boston, Birkhauser: 187–199.

"Tumors of the Connective and Supporting Tissues," Suit, H. (1995), Regaud Lecture, Granada, Radiother & Oncol.

"Adult Soft Tissue Sarcomas of the Head and Neck Treated by Radiation and Surgery or Radiation Alone: Patterns of Failure and Prognostic Factors." Willers, H., E. Hug, et al. (1995), Int. J. Radiation Onc., Biol., Phys. 33(3):585–593.

"Production Cross Sections for C^{14} from Elements Found in Lunar Rocks; Implications for Cosmic Ray Studies [Abstract]," Sisterson, J., R. J. Schneider, et al. (1995), Lunar and Planetary Science XXVI, Houston, TX.

"Facilities Under Construction, Planned and Proposed," Sisterson, J. M. (1995), in *Ion Beams in Tumor Therapy*. U. Linz. Weinheim, Chapman and Hall: 371–381.

"Getting Started: The First Years of Biomedical Protons. [Abstract]," Koehler, A. M. (1995), XXIII Meeting of PTCOG, Cape Town, South Africa, Proceedings.

"Comparative Treatment Planning for Nasopharynx Tumors [Abstract]," Smith, A., J. Adams, et al. (1995), XXIII Meeting PTCOG, Cape Town, South Africa, Proceedings.

"Hyperfractionated, Accelerated Radiation Therapy of Advanced Paranasal Sinus Carcinoma Employing Combined Proton and Photon Irradiation [Abstract]," Thornton, A. F., M. Joseph, et al. (1995), XXIII Meeting of PTCOG, Cape Town, South Africa, Proceedings.

"Cross sections Needed for the Interpretation of Long-lived and Short-lived Cosmogenic Nuclide Production in Extraterrestrial Materials [Abstract]," Sisterson, J. M., A. Beverding, et al. (1995), 58th Annual Meeting of the Meteoritical Society, Washington, DC, Meteoritics.

"*Radiation Therapy Physics*." Smith, A. R., Ed. (1995), Heidelberg, Germany, Springer Verlag.

"Management of Anterior Cranial Base and Cavernous Sinus Neoplasms with Conservative Surgery Alone or in Combination with Fractionated Photon or Stereotactic Proton Radiotherapy," Ojemann, R. G., A. F. Thornton, et al. (1995). Clinical Neurosurgery – Proceedings of the Congress of Neurological Surgeons, Chicago, Illinois, 1994, Williams & Wilkins, Baltimore.

"Re-evaluation of Large AVM Complication Rates," Chapman, P., A. F. Thornton, et al. (1995), J Neurosurgery 82:1095–7.

1996

"Intensity Modulated Therapy And Inhomogeneous Dose to the Tumor: A Note of Caution [Editorial]," Goitein, M. and A. Niemierko (1996), Int J Radiat Oncol Biol Phys 36(2):519–522.

"Assessment of the Impact of Local Control on Clinical Outcome," Suit, H. (1996), Front Radiat Ther Oncol. Basel, Karger. 29.

"Physical Specifications Of Clinical Proton Beams from a Synchrotron," Arduini, G., R. Cambria, et al. (1996), Med Phys 23:939–951.

"Dosimetric Results from a Feasibility Study of a Novel Radiosurgical Source for Irradiation of Intracranial Metastases," Douglas, R., J. Beatty, et al. (1996), Int J Radiat Oncol Biol Phys 36 (2):443–450.

"Field Edge Smoothing for Mulitleaf Collimators," Galvin, J., D. Leavitt, et al. (1996), Int J Radiat Oncol Biol Phys 35(1):89–94.

"Conformal Irradiation of the Prostate: Estimating Long-term Rectal Bleeding Risk Using Dose-volume Histograms," Hartford, A., A. Niemierko, et al. (1996), Int J Radiat Oncol Biol Phys 36(3):721–730.

"A Pencil Beam Algorithm for Proton Dose Calculations," Hong, L., M. Goitein, et al. (1996), Phys Med Biol 41:1305–1330.

"Biological Modeling in Treatment Planning," Niemierko, A. (1996), 15th European Society for Therapeutic Radiology and Oncology Annual Meeting, Vienna, Austria.

"Visual Field Deficits Associated with Proton Beam Irradiation for Parapapillary Choroidal Melanoma," Park, S., S. Walsh, et al. (1996),Ophthalmology 103(1):110–116.

"Cauda Equina Tolerance to Radiation Therapy," Pieters, R., D. O'Farrell, et al. (1996), Int J Radiat Oncol Biol Phys 36(S1):359.

"Relationship between Dose to Auditory Pathways and Audiological Outcomes in Skull Base Tumor Patients Receiving High Dose Proton/Photon Radiotherapy," Schoenthaler, R., B. Fullerton, et al. (1996), Int J Radiat Oncol Biol Phys 36:291.

"Sarcomas of Bone and Soft Tissue," Suit, H., H. Mankin, et al. (1996), Cancer Manual, American Cancer Society Massachusetts Division: 482–94.

"Constitutional Alterations in P^{16} in Patients with Uveal Melanomas," Wang, X., K. Egan, et al. (1996), Melanoma Research 6(6):405–410.

"Time to Second Prostate Specific Antigen Failure is a Surrogate Endpoint for Prostate Cancer Death in a Prospective Trial of Therapy for Localized Disease," Zietman, A., K. Dallow, et al. (1996), Urology 47:236–239.

"Short-lived Cosmogenic Radionuclide Production in Lunar Rocks; Improved Estimates for the Solar Proton Flux in Recent Solar Cycles [Abstract]," Sisterson, J. M., R. J. Schneider-IV, et al. (1996), Lunar and Planetary Science XXVII:1207–8.

"Present and Future of the Clinical Use of Radiobiological Models," Niemierko, A. (1996), 5th International Conference on Applications of Physics in Medicine and Biology, Trieste, Italy.

1997

"Successful Treatment of Esthesioneuroblasatoma and Neuroendocrine Carcinoma with Combined Chemotherapy and Proton Radiation," Bhattacharyya, N., A. Thornton, et al. (1997), Arch Otolaryngol Head Neck Surg (123):34–40.

"Brainstem Tolerance to Conformal Radiotherapy of Skull Base Tumors," Debus, J., E. Hug, et al. (1997), Int J Radiat Oncol Biol Phys 39(5):967–975.

"Results of a Phase II Study with Proton/Photon Irradiation to 90 Cobalt Gray Equivalent (CGE) in Accelerated Fractionation for Glioblastoma Multiforme," Fitzek, M., A. Thornton, et al. (1997), Int J Radiat Oncol Biol Phys 39(S2).

"Neuropsychological Function in Adults after High Dose Fractionated Radiation Therapy of Skull Base Tumors," Glosser, G., P. McManus, et al. (1997), Int J Radiat Oncol Biol Phys 38(2):231–239.

"The Probability of Controlling an Inhomogeneously Irradiated Tumor: A Stratagem for Improving Tumor Control through Partial Tumor Boosting, Goitein, M., A. Niemierko, et al. (1997), 19th LH Gray Conference.

"The technology of Hadrontherapy: The Context within which Technical Choices are Made," Goitein, M. (1997), U. Amaldi, B. Larsson and Y. Lemoigne, Advances in Hadrontherapy, Elsevier Science B.V.:141–159.

"1996 Jules Gonin Lecture of the Retina Research Foundation. Long-term Results after Proton Irradiation of Uveal Melanomas," Gragoudas, E. (1997), Graefe's Arch Clin Exp Ophthalmol 235(5):265–267.

"Optic Neuropathy Following Combined Proton and Photon Radiotherapy for Base of Skull Tumors. [Abstract]," Kim, J., J. Munzenrider, et al. (1997), 39th Annual ASTRO Meeting, Int J Radiat Oncol Biol Phys.

"Gangliogliomas in Adults," Hakim, R., J. Loeffler, et al. (1997), Cancer 79:127–131.

"An Evaluation of Tumor Vascularity as a Prognostic Indicator in Uveal Melanoma." Lane, A., K. Egan, et al. (1997), Melanoma Research 7(3):237–242.

"The Potential Role of Proton Beams in Radiation Oncology," Loeffler, J., A. Smith, et al. (1997), Seminars in Oncology 24(6):686–695.

"Radiation Tolerance of the Cervical Spinal Cord: Incidence and Dose-Volume Relationship of Symptomatic and Asymptomatic Late Effects Following High Dose Irradiation of Paraspinal Tumors [Abstract]," Liu, M., J. Munzenrider, et al. (1997), 39th Annual ASTRO Meeting, Int J Radiat Oncol Biol Phys.

"Patient Assessment, Treatment Planning And Dose Delivery," Lomax, A. and M. Goitein (1997), Advances in Hadrontherapy. U. Amaldi, B. Larsson and Y. Lemoinge, Elsevier Science B.V.: 233–250.

"Kerma Measurements in Polyenergetic Neutron Fields," Newhauser, W., U. Schrewe, et al. (1997), Radiation Protection Dosimetry 70(1–4):13–16.

1998

"Primary and Metastatic Brain Tumors," DeAngelis, L., J. Loeffler, et al. (1998), in *Cancer Management: A Multidisciplinary Approach.* L. C. R Pazdur, WJ Hoskins, LD Wagman. NY, PRR, Hungington: 1–21.

"Eye preservation by combined surgery and conformal proton irradiation for malignant tumors of the orbit," Fitzek, M., A. Thornton, et al. (1998), Int J Radiat Oncol Biol Phys 42(S1):330.

"Radiosurgery. Textbook of Radiation Oncology," Larson, D., D. Shrieve, et al. (1998), S. Leibel and T. Phillips. Philadelphia, WB Saunders: 383–399.

"Fractionated Stereotactic Radiosurgery: Lessons from Radiosurgery," Loeffler, J. and D. Shrieve (1998), Progress in Radio-Oncology VI. H. Kogelnik and F. Sedlmayer. Bologna, Monduzzi Editore. VI: 75–88.

"Present and Future of the Clinical Use of Radiobiological Models In Radiation Therapy Treatment Planning," Niemierko, A. (1998), Medica Physica 13(S1):34–38.

"Potential Clinical Impact of Normal-tissue Intrinsic Radiosensitivity Testing [Letter to the Editor]," MacKay, R. I., A. Niemierko, et al. (1998), Radiotherapy and Oncology 46(2):215–216.

"Radiobiological Models of Tissue Response To Radiation In Treatment Planning Systems," Niemierko, A. (1998), Tumori 84(2):140–144.

"Monte Carlo Method to Study the Proton Fluence For Treatment Planning," Paganetti, H. (1998), Med Phys 25(12):1–6.

"Temporal Lobe Damage Following Surgery and High Dose Photon and Proton Irradiation in 96 Patients Affected by Chordomas and Chondrosarcomas of the Base of Skull," Santoni, R., N. Liebsch, et al. (1998), Int J Radiat Oncol Biol Phys 41(1):59–68.

"Interstitial Brachytherapy for the Treatment of Brain Metastases," Schulder, M. and J. Loeffler (1998), Advenced Techniques in Central Nervous System Metastases. R. Maciunas, Neurosurgical Topics of the American Association of Neurological Surgeons: 155–164.

"Accelerated, Hyperfractionated Proton/Photon Irradiation for Advanced Paranasal Sinus Cancer: Results of a Prospective Phase I-II Study," Thornton, A., M. Fitzek, et al. (1998), Int J Radiat Oncol Biol Phys 42(S1):222.

"Benign Meningioma Partially Resected and Recurrent Intracranial Tumors Treated with Combined Proton and Photon Radiotherapy," Wenkel, E., A. Thornton, et al. (1998), Int. J Radiat Oncol Biol Phys 42(S1):271.

"C^{14} Depth Profiles in Apollo 15 and 17 Ores and Lunar Rock 68815," Jull, A. J. T., S. Cloudt, et al. (1998), Geochim. Cosmochim. Acta. 63:3025–36.

"Status of Production Cross Section Measurements Needed for Cosmic Ray Studies [Abstract]," Sisterson, J. M. (1998), Lunar and Planetary Science XXIX. Abstract 1189, Lunar and Planetary Institute, Houston.

"Neon Production from Al, Mg, Si, and Fe: New Proton Production Cross Section Measurements [Abstract]," Sisterson, J. M. and M. W. Caffee (1998), Lunar and Planetary Science XXIX. Abstract 1234, Lunar and Planetary Institute, Houston, (CD-ROM avalailable).

"Survival Implications of Enucleation after Difinitive Radiotherapy for Choroidal Melanoma: An Example of Regression on Time-dependent Covariates," Egan, K. M., L. Ryan, et al. (1998), Arch Ophthalmol 116:366–370.

"High Performance Medical Robot Requirements and Accuracy Analysis." Mavroidis, C., J. Flanz, et al. (1998), Robotics and Computer-Integrated Manufacturing 14:329–338.

"Calculation of the Spatial Variation of the Relative Biological Effectiveness in a Therapeutic Proton Field for Eye Treatment," Paganetti, H. (1998), Phys. Med. Biol 43:2147–57.

"Skull Base Tumors: Treatment with Three-dimensional Planning and Fractionated X-ray and Proton Radiotherapy," Munzenrider, J. E., J. Adams, et al. (1998), in *Textbook of Radiation Oncology*, S. A. Liebel and T. L. Phillips. Philadelphia, W. B. Saunders.

1999

"Stereotactic Proton Radiosurgery," Harsh G, Loeffler JS, Thornton A, Smith A, Bussiere M, Chapman P, in: Konziolka D (ed), *Applications of Radiosurgery, Neurosurgery Clinics of North America*, W.B. Saunders, Philadelphia, pp. 243–256, (1999).

"World-wide Proton Therapy Experience in 1997," Sisterson, J. M. (1999), 15th Conference on the Application of Accelerators in Research and Industry, AIP Press.

"Synchrocyclotron Survivor to Bow Out after 50 Years," Gottschalk, B., A. M. Koehler, et al. (1999), CERN Courier: 22–25.

"Proton Beam Therapy for Age-related Macular Degeneration: Development of a Standard Plan." Adams, J. A., K. L. Paiva, et al. (1999), Medical Dosimetry 24:233–238.

"Verification of the Alignment of a Therapeutic Radiation Beam Relative to its Patient Positioner," Barkhof, J., G. Schut, et al. (1999), Med. Phys. 26:2429–37.

"History of Childbearing Associated with Improved Survival in Choroidal Melanoma." Egan, K. M., J. L. Quinn, et al. (1999), Arch Ophthalmol 117:939–42.

"Accelerated Fractionated Proton/Photon Irradiation to 90 Cobalt Gray Equivalent for Glioblastoma Multiforme." Fitzek, M. M., A. F. Thornton, et al. (1999), J. Neurosurgery 91:251–60.

"Recent Performance of the NPTC Equipment Compared with the Clinical Specifications," Flanz, J. B., J. Bailey, et al. (1999), Conference on the Application of Accelerators in Research and Industry, AIP Press.

"Initial Equipment Commissioning of the Northeast Proton Therapy Center," Flanz, J. B., S. G. Bradley, et al. (1999), Cyclotron Conference, IOP Publishing.

"Relative Biological Effectiveness of Proton Beams in Clinical Therapy," Gerweck, L. E. and S. V. Kozin (1999), Radiother. Oncol. 50:135–42.

"Nuclear Interactions of 160 MeV Protons Stopping in Copper: A Test of Monte Carlo Nuclear Models." Gottschalk, B., R. Platais, et al. (1999), Med Phys 26:2597–2601.

"Risk factors for Radiation Maculopathy and Papillopathy after Intraocular Irradiation," Gragoudas, E. S., W. Li, et al. (1999), Ophthmal. 106:1571–8.

"Sterotactic Proton Radiosurgery," Harsh, G., J. Loeffler, et al. (1999), Neurosurgery Clinics of America. D. Konziolka. Philadelphia, W. B. Saunders. 20:243–56.

"Proton Radiotherapy," Hartford, A. C., A. L. Zietman, et al. (1999), in Radiothereutic Management of Carcinoma of the Prostate, A. D'Amico and G. E. Hanks. London, UK, Arnold Publishers: 61–72.

"Proton Therapy for Uveal Melanomas and Other Eye Lesions," Munzenrider, J. E. (1999), Strahlenther. Onkol. 175 (Suppl. 2):68–73.

"Proton Therapy for Tumors of the Skull Base," Munzenrider, J. E. and N. J. Liebsch (1999), Strahlenther. Onkol. 175 (Suppl. 2):57–63.

"Clinical Use of Particle Radiation Therapy," Munzenrider, J. E. (1999), in Clinical Radiation Oncology: Indications, Techniques and Results. C. C. Wang, John Wiley and Sons.

"Iris Color as a Prognostic Indicator in Intraocular Melanoma," Regan, S., K. M. Egan, et al. (1999), Arch Ophthalmol 117:811–4.

"Chondrosarcoma of the Base of the Skull: A Clinicopathologic Study of 200 Cases with Emphasis on its Distinction from Chordoma," Rosenberg, A. E., G. P. Nielson, et al. (1999), Am. J. Surg. Path. 23:1370–8.

"Germline BRCA2 Sequence Variants in Patients with Ocular Melanoma," Serova-Sinilnikova, O., K. M. Egan, et al. (1999), Int J Cancer 82:1571–8.

"Analysis of the Relationship between Tumor Dose Inhomogeneity and Local Control in Patients with Skull Base Chordoma," Terahara, A., A. Niemierko, et al. (1999), Int J Radiat Oncol Biol Phys 45:351–8.

"Proton Dosimetry Intercomparison Based on the ICRU 59 Protocol." Vatnitsky, S., M. Moyers, et al. (1999), Radiother Oncol 51:273–9.

2000

"Proton Beam Therapy," Harsh GR IV, Smith AR, Loeffler JS, in Berstein M and Berger MS (eds), *Neuro-oncology, The Essentials,* New York, Thieme Medical, pp. 210–217, (2000).

"Proton-beam Radiation Therapy," Loeffler JS, Singer RJ, Chapman PH, Ogilvy CS, in Germano IM (eds), *LINAC and Gamma Knife Radiosurgery,* Park Ridge (IL), American Association of Neurological Surgeons, pp. 71–75, (2000).

"The measurement of Proton Stopping Power using Proton-cone-beam Computed Tomography," Zygmanski, P., Gall, K.P., Rabin, M.S.Z. and Rosenthal, S.J., Phys. Med. Biol. 45:511–528 (2000).

"Randomized Controlled Trial of Varying Radiation Doses in the Treatment of Choroidal Melanoma." Gragoudas, E. S., A. M. Lane, et al. (2000), Arch Ophthalmol 118:773–8.

"Fractionated, Three-dimensional Planning-assisted Proton-radiation Therapy Rhabdomyosarcoma: A Novel Technique," Hug, E. B., J. Adams, et al. (2000), Int. J. Radiat. Oncol. Biol. Phys. 47:979–84.

"Management of aTypical and Malignant Meningiomas: Role of High Dose, 3d- Conformal Radiation Therapy," Hug, E. B., A. DeVries, et al. (2000), J Neurooncol. 48:151–60.

"Metastatic Melanoma Death by Anatomic Site after Proton Beam Irradiation for Uveal Melanoma." Li, W., E. S. Gragoudas, et al. (2000), Arch. Ophthmalmol. 118:1066–70.

"Patterns of Tumor Initiation in Choroidal Melanoma." Li, W., H. Judge, et al. (2000), Cancer Res. 60:3757–60.

"Radiobiological Significance of Beam Line Dependent Proton Energy Distributions in a Spread-out Bragg Peak," Paganetti, H. and M. Goitein (2000), Med. Phys. 72:1119–26.

"The General Relation between Tissue Response to X-radiation and the Relative Biological Effectiveness (RBE) of Protons: Prediction by the Katz Track Structure Model," Int. J. Radiat. Biol. 76:985–98.

"Benign Meningioma: Partially Resected and Recurrent Intracranial Tumors Treated with Combined Proton and Photon Radiotherapy," Wenkel, E., A. F. Thornton, et al. (2000), Int. J. Radiat. Oncol. Biol. Phys. 48:1363–70.

"The measurement of Proton Stopping Power Using Proton-cone-beam-computed-tomography." Zygmanski, P., K. P. Gall, et al. (2000), Phys Med Biol 45:511–28.

"Proton Beam Radiation Therapy," Spiro IJ, Lomax A, Smith AR, Loeffler JS, in DeVita VT, Hellman S, Rosenberg SA (Eds). *Cancer: Principles and Practice of Oncology.* 6th Edition, Lippincott-Raven, Philadelphia, pp. 3229–3235, (2001).

"Dose-escalation with Proton/Photon Irradiation for Daumas-duport Lower Grade Glioma: Results of an Institutional Phase I/II Trial," Fitzek, M. M., A. F. Thornton, et al. (2001), Int. J. Radiat. Oncol. Biol. Phys. 51:929–835.

"Biophysical Modeling of Proton Radiation Effects Based on Amorphous Track Models," Paganetti, H. and M. Goitein (2001), Int. J. Radiat. Biol. 77:911–928.

"Hypothalamic/Pituitary Gland Dysfunction Following High-dose Conformal Radiotherapy to the Base of Skull: Demonstration of a Dose-effect Relationship Using Dose-volume Histogram Analysis," Pai, H. H., A. Thornton, et al. (2001), Int. J. Radiat. Oncol.

2002

"Proton Beam Radiation Therapy." Smith AR, Loeffler JS, in Bertino JR (ed), *Encyclopedia of Cancer*, Academic Press, San Diego, CA, pp. 481–490, (2002).

"Proton Beam Therapy in the Management of Bone and Soft Tissue Sarcomas," DeLaney TF, Curr Opin Orthop 2002; 13:434–439.

"Proton Irradiations of Large Area Hg Cd Photovoltaic Detectors" M. W. Kelly, E. Ringdahl, A. I. D'Souza, S.D. Luce, J. Ehlert, and E. W. Cascio Infrared Technology and Applications XXVII, Proceedings of SPIE, (Seattle Washington) 4820, 479.

2003

"Proton Beam Radiation Therapy: Principles and Practice of Oncology Uppdates," DeLaney TF, Smith AR, Lomax A, Adams J, Loeffler JS, 17(1):1–10, 2003.

"Gamma Ray and Neutron Spectrometer for the Dawn Mission to 1 Ceres and 4 Vesta," T. H. Prettyman, W. C. Feldman, F. P. Ameduri, B. L. Barraclough, E. W.Cascio, K. R. Fuller, H. O. Funsten, D. J. Lawrence, G. W. McKinney, C. T. Russell, S. A. Soldner, S. A. Storms, C. Szeles and R. L. Tokar IEEE/TNS 50(4), Aug. 2003 (accepted for publication).

"Weighing Proton Therapy's Clinical Readiness and Costs" B.Gottschalk, A. M. Koehler, and R. Wilson, Physics Today (letter) 56(6):10 (2003).

Index

Val Kirsis, 84
Xiaowei Yan, 46
Yetin Orshen, 46